Creative Industries in Syria

Changes and Adaptation

Foreword by Sarab Atassi, PhD

Participating researchers: Dima Nachawi | Ola Alshikh Hassan | Zeina Shahla

These researches were carried out with the support of Ettijahat–Independent Culture as part of the fifth edition of the programme "Research to Strengthen the Culture of Knowledge" in 2018.

CREATIVE INDUSTRIES IN SYRIA

Changes and Adaptation

Foreword by Sarab Atassi, PhD

Participating researchers: Dima Nachawi | Ola Alshikh Hassan | Zeina Shahla

ibidem-Verlag
Stuttgart

Bibliografische Information der Deutschen Nationalbibliothek
Die Deutsche Nationalbibliothek verzeichnet diese Publikation in der Deutschen Nationalbibliografie; detaillierte bibliografische Daten sind im Internet über http://dnb.d-nb.de abrufbar.

Bibliographic information published by the Deutsche Nationalbibliothek
Die Deutsche Nationalbibliothek lists this publication in the Deutsche Nationalbibliografie; detailed bibliographic data are available in the Internet at http://dnb.d-nb.de.

Proof-reading: Charles Huckabee
Cover design: Ibrahim Brimo
Foreword by Sarab Atassi, PhD
Translator: Gilgamesh Nabeel

Ettijahat-Independent Culture
Main office: Boulevard Louis Schmidt 119/2,1040,
Bruxelles, Belgique
Phone: 00961-1-442770
Email: info@ettijahat.org
Website: www.ettijahat.org

Opinions explained within these research papers do not necessarily reflect *Ettijahat*'s opinion nor its members.

Cover image: Birds, Hussein Madi, Steel, Courtesy of the Atassi Foundation for Arts and Culture

Gedruckt auf alterungsbeständigem, säurefreien Papier
Printed on acid-free paper

ISBN: 978-3-8382-1401-6

© *ibidem*-Verlag
Stuttgart 2020

Alle Rechte vorbehalten

Das Werk einschließlich aller seiner Teile ist urheberrechtlich geschützt. Jede Verwertung außerhalb der engen Grenzen des Urheberrechtsgesetzes ist ohne Zustimmung des Verlages unzulässig und strafbar. Dies gilt insbesondere für Vervielfältigungen, Übersetzungen, Mikroverfilmungen und elektronische Speicherformen sowie die Einspeicherung und Verarbeitung in elektronischen Systemen.

All rights reserved. No part of this publication may be reproduced, stored in or introduced into a retrieval system, or transmitted, in any form, or by any means (electronic, mechanical, photocopying, recording or otherwise) without the prior written permission of the publisher. Any person who does any unauthorized act in relation to this publication may be liable to criminal prosecution and civil claims for damages.

Printed in the EU

Research
To Strengthen the Culture of Knowledge

This book was produced in the framework of the fifth edition of the programme Research: To Strengthen the Culture of Knowledge. This capacity building programme seeks to support Syrian and Palestinian Syrian cultural actors between 22 and 40 years of age, to provide them with skills and knowledge in the field of cultural research. In addition to providing resources to those interested in this field, the programme provides young researchers with the opportunity to carry out a research project (which may be their first research outside the framework of the academic study) under the direct supervision of researchers specialized in the field of culture. The programme also supports them in showcasing the results of their research papers and discussing them on an academic level that will play a role in enriching their research content.

The programme focuses on current topics in cultural research that are relevant to the situation in Syria, especially changes in the perspectives of Syrian artists with regard to their relationship with society and the transformations it is witnessing, resulting in a transformation in the role of culture and arts in Syria in the near future.

The fifth edition of the programme's duration was 9 months consisting of three key phases:

1) The application submission and the selection of young researchers: 2 months
2) Training and skill enhancement phase: 1 month
3) Completion of actual research supervised by experienced researchers: 6 months

At the end of the programme, and based on the evaluation of an expert committee, Ettijahat publishes a collection of the completed papers noting that this activity is based on the evaluation of all the papers equally.

Ettijahat-Independent Culture would like to express its gratitude to all its partners for their support. Special thanks go to Mrs. Laila Hourani and Ford Foundation, Turquoise Mountain, and Mimeta – Centre for Culture Sector Development and Arts Cooperation for their support since the beginning of the programme and for their faith in the value of cultural research.

We owe our utmost recognition to all the respectable trainers, lecturers, scientific committee members and mentors with whom we have we had the privilege of collaborating, particularly for their efforts, involvements and commitments. Special mention in this regard is due to Hassan Abbas, PhD, and Marianne Njeim, PhD.

The programme would not have had the same impression were it not for the involvement of the young researchers, who have shown genuine enthusiasm and seriousness in conducting research on topics related to their country, communities and fields of work, and from whom we – as a team – have learned a great deal.

www.ettijahat.org
Facebook: Ettijahat-Independent Culture
Twitter: EttijahatIndep
Youtube: Ettijahat-Independent Culture

With the kind support of:

Contents

Foreword **11**

The Impact of War on the Aghabani Textile Industry **21**
    Summary 22
    Introduction 24
    Definitions and Terminologies 35
    Literature Review 37
    Part I: The Historical Context of the Aghabani Industry and the Changing Dynamics of Its Work After 2011 43
    Part II: Response Mechanisms to the Risks and Challenges Facing the Aghabani Industry 63
    Conclusion 87
    References 92
    Appendix 101

Changes in the Gender Roles of Women Handicraft Workers in Damascus After 2011 .... **105**
    Summary 106
    Introduction 108
    Part I: The Economic and Social Conditions of Female Handicraft Workers in Damascus and Its Countryside 114
    Part II: Monitoring the Gender Variables 124
    Conclusion 137
    References 139
    Appendix 141

The Damascene Textile Industry in the Cauldron of War **145**
    Acknowledgements 146
    Summary 147
    Introduction 150
    Part I: Damascene Textile Crafts from Ancient Times until 2011 156

Part II: The Reality of the Damascene Fabrics in the Shadow of War 178

Conclusion and Recommendations 202

References 207

Appendix 1: Proverbs and Folk Songs Related to the Textile Crafts 214

Appendix 2: Quantitative Table of Some of the Study's Results 216

Foreword

Sarab Atassi, PhD
July 2019

"Cultural heritage is a mirror of the community's life, history and identity. Preserving it helps in rebuilding divided societies, restoring their identity and connecting their past with their present and future."—UNESCO

"Ettijahat–Independent Culture" was born in one of the darkest periods in the Arab region in general and in Syria in particular. What happened to Syria since 2011 is a devastating catastrophe for both people and place. Ettijahat has chosen to be one of the bright windows amid such darkness and to dedicate its programme to activating independent cultural work to be more positive and influential in the process of cultural, political and social change in Syria.

In 2014, after conducting in its third year an in-depth participatory research work, Ettijahat published a consensus document on the priorities of cultural action in Syria. A proposal was put forward for the role of culture and arts at the current stage and in the near future, along with efforts to build a consensus ground among independent actors and artists that would allow the development of general trends as an alternative cultural policy to cope with Syria's profound transition.

Since its beginning, Ettijahat has set its main work along several axes that it actively pursues, cooperating with a number of specialists in the field of culture. It has consistently advanced its work along these axes through, in its own words, "supporting artists and cultural initiatives, empowering young researchers, building agreements and alliances among individuals and cultural institutions, and promoting the arts and artists through regional and international platforms, in addition to working to make culture and art accessible to the Syrian communities wherever they are."

Ettijahat's most important and positive aspect is its ongoing effort to establish young scholars in the practice of critical scientific methodology: something the organization takes special and continuing care of, especially since we, the Syrians, have lacked such a fundamental feature in our educational institutions in a way that is no longer acceptable, and especially in comparison to countries we were in the forefront of a few decades ago.

This effort is evident through the follow-up of a programme entitled "Research: to Strengthen the Culture of Knowledge", which involves the empowerment of Syrian and Syrian-Palestinian young researchers. This programme has contributed to a number of important field and academic studies. In 2019, the programme will conclude its sixth edition. During these six editions, more than sixty-five male and female scholars have been supported in research projects. Some of their research papers were also published on Ettijahat's website. In 2016, five research papers were selected from the second edition (2014) and were published in a book of 334 pages presented by Dr. Hassan Abbas. Four other papers were published in 2018 and presented by Dr. Jamal Chehayed. Reviewing these results, we recognise the programme's importance and its scientific and objective approach to the current Syrian situation.

In this context, five research papers were selected from the fifth edition for publication. This is an introduction to three of these integrated and closely related research topics, among the studies completed in the fifth edition of 2018.

The fifth edition was decided on after the programme proved the line it was following through the diversity and seriousness of the studies conducted during the first four editions, and indeed Ettijahat chose to expand the cultural and artistic fields by adding two axes that would widen the horizons of participation by linking two main issues related to heritage and documentation. Anyone who has navigated in these two fields knows very well their importance.

The topics of the first years were:

- Studying the cultural structures active in Syrian society and researching the causes of cultural marginalization, besides studying the dynamics of transformation in the features of Syrian culture and its inherited characteristics during the civil protest movement and its various stages.

- Studying the features of the new Syrian culture generated during the protest movement and exploring its prospects and possibilities of development after the crisis.
- Studying the artistic and creative trends of Syrian art products and Syrian artists wherever they are.
- Studying the management of the cultural change process and the possible forms of intervention.

The two new topics added in the fifth edition were:

- Researching the creative crafts industries, including the materials, production techniques and the related professional expertise.
- Exploring forms of documenting, archiving and protecting heritage, traditional or contemporary, and exploring the role of this heritage in the social, creative and economic sectors of Syria in the future.

Over these five years, the research topics have varied. With a quick glance at the research papers' titles, we appreciate the importance of these theoretical and field studies, which constitute a reference and future pillar as they followed with precision and objectivity the transformations of the current situation.

The added focus in the programme's fifth edition, on traditional craftsmanship and documentation, is very important and allows for a re-evaluation of the current status of this knowledge. The scholars have studied in depth some trades that need accurate fieldwork and direct connection with the street, labour market and human reality. We also noticed the applications of young men and women from inside Syria and their proposals to conduct important research based on field studies. Their presence in the programme, along with the colleagues who had left Syria, gave them opportunities to open up to different experiences in an atmosphere of freedom of expression.

Protecting tangible and intangible cultural heritage, along with exploring the forms of documentation and archiving, are also priorities of UNESCO's "Emergency Safeguarding of the Syrian Cultural Heritage" programme, which has been intensively active in Beirut during the course of the war. In its scientific meetings, training courses and other activities, there was an emphasis on the preservation of documents, copying all their types. Eventually, the topic of intangible cultural heritage got a good share of attention for being quite related to the issue of protection of memory and identity in the midst of war and destruction.

The work of this programme was important as it facilitated opportunities for concerned stakeholders and cultural actors to meet up in Beirut, in addition to holding some workshops and training sessions on practical applications. It also allowed for setting priorities while providing opportunities for scientific dialogue and meetings between concerned actors in these fields. This general briefing stressed the importance of the urgent preservation of the exceptional Syrian heritage in a war that has harmed people, monuments, heritage and history. However, this protection by a responsible international body does not dispense with the need for in-depth research work that is based on the persistent efforts of local scientists, researchers, experts and technicians. Hence, in the darkness of the current horizon, the importance of Ettijahat's Research programme is evident, namely the scientific establishment of young Syrian researchers to form the nucleus of those who will follow this path in the field of heritage, which is one of the pillars of broad cultural interest.

Since the beginning of the conflict and amidst the surrounding destruction, I joined the UNESCO programme as a scholar specialised in documenting and protecting the old city of Damascus. Within this framework, which focused primarily on the protection of monuments and archaeological sites, museums and antiques, I was thinking about the protection of historical Syrian cities and the importance of this process that was taking place under complicated humanitarian conditions. Here, the priorities have crystallised on the subject of memory, the memory of people and places (including the memory of traditional crafts industries), and the importance of this work in maintaining cohesion and association, and its priority in the protection of national identity. This work requires the experience and the combined efforts of three generations: one holds the distant memory, while the next has scientific knowledge and field experience prior to the war years and has the specialties of the applications and essence of history, architecture, environment, social sciences, as well as arts; and a third generation carries the banner of modern scientific follow-up methodology. The last is a young generation that understands the dimensions of the Syrian conflict, documents the present and paves the way for the future with regard to the threatened tangible and intangible cognitive heritage, including the traditional crafts (working with textiles, embroidery, wood, glass, etc.).

Impressive and promising are the research papers we present to you in this section of the book, which has been conducted and presented in a short time and at an excellent scientific level by three young Syrian women who have dealt in depth with the selected topics, which dealt with some of the ancient Syrian heritage related to the textile and embroidery trade, and the labour force that works on and supervises the manufacturing and marketing of these crafts without forgetting the current situation and its repercussions.

These studies formed a new step in the research programme, and Ettijahat might be able to find a supportive framework to follow up this line. The stage is critical for preserving the memory of these professions and the possibility of benefiting from those who persevered in them and continued to work in Damascus at the beginning of the current situation, despite the destruction of the vast green and rural areas that surround the city and form its vital dimension and framework in many aspects, including traditional handicrafts.

Other Syrian cities, such as Homs, Hama, Lattakia, and especially Aleppo, which continues to devote most of its efforts to large-scale emergency operations related to its destroyed architectural heritage, should not be neglected in this regard. Aleppo's markets, its traditional industries, and their owners and people are in the core of the rehabilitation efforts of the old city, which was severely damaged.

The three selected research papers are as follows:

- "The Impact of War on the Aghabani Textile Industry (Damascus as a Case Study)", by Dima Nachawi

This research presents a rich addition related to the intangible heritage aspect of the Aghabani industry by collecting and analysing information on the history and conditions of the industry, and participating in a field study in Damascus as a case study. In this study, we read about the motivations of all individuals in the research sample and their adherence to the Aghabani industry because of its association with their identity. This is a form of cultural resistance and a means of protecting a distinctive and outstanding industry that was concentrated in the city of Douma at the gates of Damascus.

The importance of this paper lies in its taking note of the administrative patterns adopted in Syria during the current time and the new initiatives to support traditional industries and intangible cultural heritage. Through this example of the Aghabani industry and its marketing, we can notice the great negligence that still surrounds the rights of workers and their protection in exchange for the tangible potentials for the development of this artistic Damascene heritage. This research also presents a fundamental argument concerning the ability of the cultural management of traditional handicrafts to provide a balance between cultural and aesthetic standards on the one hand and, on the other, the material concept resulting from the sale of culture as a commodity that generates material profit without neglecting the rights of the artistic labour force.

- "Changes in the Gender Roles of Women Handicraft Workers in Damascus After 2011: An Exploratory Paper", by Ola Alshikh Hassan

This study noted the changes within a sample of Syrian women whose circumstances changed after 2011 due to the ongoing war in Syria within Damascus and its countryside. Through extensive in-depth interviews and questionnaires, the study attempted to answer a key question: Have the women who entered the handicrafts market in Damascus and its countryside become agents of change after most families were displaced and forced to abandon their homes, with the husband often missing or absent? In the end, Ola el-Sheikh Hassan raises the question of the usefulness of these changes, for it drives Syrian women to take care of public affairs and encourages them in the longer term towards practising real citizenship.

- "The Damascene Textile Industry in the Cauldron of War", by Zeina Shahla

In this research, three Damascene textile handicrafts were chosen, namely brocade, al-Aghabani, and handmade rugs. The study began with a question about what changes had taken place in Damascus' textile handicrafts, and whether these crafts had become endangered due to the war circumstances. The study included the current reality of these crafts and compared that with how they were before the war. The research community included samples of crafts practitioners of all levels, fields and stages of production, in addition to owners of shops selling Damascene cloth products and the relevant governmental bodies (the Union of Craftsmen, Union of Craft Associations and Industrial Secondary Schools). Zeina Shahla's study confirmed that the labour force is the most affected part of the textile handicrafts industry and the one posing the greatest threat to the possibility of continuing such trades. The human resource pool in Syria has been subjected to unprecedented bleeding as a result of migration, displacement, deaths and injuries, and this forms a great challenge for workers in these trades and those who carry them.

Today, after completing their valuable papers, Zeina Shahla consciously continues observing the cultural and artistic life in Damascus and its environs through her beautiful, powerful writings; Dima Nachawi is still working and has become more brilliant in her contribution in more than one humanistic artistic field, not to forget her potentials and competence in the field of social research in association with studying creative and historical artistic crafts; while Ola Alshikh Hassan is engaged on a daily basis in the heart of the hope industry along with women, the Aghabani industry and embroidery. This young woman has put her hand on the wound by addressing the issue of Syrian women. Yes, it is true that the Syrian society is exhausted today and has a clear majority of women, and therefore the women of Syria will carry an important part of the responsibility and the difficulty of rebuilding the country in a difficult economic and social reality. They will also have a role in how to rebuild the social networks and in the end they must have a presence in the making of Syria after the war.

The three scholars brought to their research a strong belief in the connection between heritage and national identity as well as a continuous follow-up to the country's conditions, with a deep human sense and understanding of the scale of the conflict in Syria. The importance of these studies is due to the originality of the presented crafts and their connection to the Damascene and Syrian identity. The researchers also note the seriousness of what the reality of these crafts might be amidst the lack of organised and systematic efforts to document and preserve them while also developing them, in addition to supporting the rights of professionals working in these traditional industries. Of course, it is difficult to grasp the reality of all the Damascene crafts within the boundaries of these papers. However, the selected models represent a good sample, many of whose features can be generalised to other crafts, bearing in mind the specificity of each one. The suffering is one and the same, and the steps that must be taken urgently to protect this important part of our heritage from loss are also similar for all traditional crafts. Such efforts must include all of them without exception.

In the end, we must return to the recent history of documenting the traditional industries and arts. Here, we remember the role of Mr. Shafiq al-Imam and his passion in this area, which he entered alone with a great sense of responsibility in the late 1940s, when he founded Syria's first museum of the arts, customs and folk traditions in al-Azm Palace, one of Damascus' most beautiful houses. There began the work of collecting, documenting and recording a variety of examples of traditional arts and industries, of which a part can be seen upon wandering in the corners and halls of the palace.

I met Shafiq al-Imam in the beginning of 1970s while we were fighting the acceleration of the changes in the old city of Damascus, amid a lack of systematic documentation of many of its parts. At the time, the idea of cooperation with the French Institute and the Imam was developed to study the traditional industries. Then, we completed an extensive study of the traditional glass industry in cooperation with the Abu Ahmed factory, a well-known factory outside Bab Sharqi, the historic Eastern Gate of Damascus. That paper has been published in French while the Arabic translation is awaiting publication. During that period, the study team continued the preparation of a second research on the manufacture of fine woods with mother-of-pearl inlay and the traditional furniture industry, and the document is still awaiting the initiative of a scholar to complete the study and follow up on the status quo.

The years passed and Shafiq al-Imam passed away on September 13, 1993, and we stopped following the project and the paper was left abandoned. Scientific exploration of the subject of arts and industries has been a slow process. For example, Professor Mounir Kayyal worked on a book he developed over a period of two decades in order to put a revised edition into our hands in 2007.

However, the model presented today by Ettijahat's "Research: to Strengthen the Culture of Knowledge" programme is similar and closer to the integrated research I mentioned on the traditional glass industry (1973–1975), which was carried out by a team of five specialised researchers. This comparison is a deep appreciation of the work of the three scholars and the rigorous and systematic supervision of "Research: to Strengthen the Culture of Knowledge" programme within the framework of Ettijahat– Independent Culture, and it is also an invitation to follow up the work in this direction, with emphasis on the need for the original Arabic version of these studies along with the translation in your hands.

The Impact of War on the Aghabani Textile Industry

Damascus as a Case Study

By: Dima Nachawi

Under the supervision of Hassan Abbas, PhD

Summary

War constitutes a real threat to the intangible cultural heritage represented by traditional handicraft industries, including the Aghabani textile industry. Although the craft still exists in Damascus, its production conditions have been changed as a result of destruction of the industry's spatial environment, the displacement or death of skilled artisans, a lack of human resources, the difficulty of obtaining raw materials, and the challenges of marketing and sales.

This research addresses the questions of whether male and female workers in the Aghabani industry are motivated to find mechanisms and solutions that would contribute to the preservation of the industry from extinction; whether the emerging challenges may have contributed to enhancing their heritage value, as the industry is part of their intangible cultural heritage, which is a reflection of their cultural identity; and whether carrying on the craft under these motivations can be classified as an act of cultural resistance.

The importance of this research is that it is one of the few studies dealing with the Aghabani industry in general and in the period between 2011 and 2018 in particular. It seeks to fill the knowledge gap about the history and conditions of this industry, analysing it and engaging in field study in Damascus, as a case study, to find solutions and recommendations from the actors themselves on how to maintain this industry.

The research focuses on gauging the impact of war on the motivation of artisans and other industry stakeholders to continue work and production, collecting the responses of concerned governmental and nongovernmental institutions, and recommending new solutions to overcome obstacles and threats to the industry. These considerations are in addition to the research's documentary goal regarding the history and development of the industry before and after 2011.

The research follows a qualitative methodology and semi-structured in-depth interviews to collect information from merchants and craftsmen in Damascus. The research sample is a non-random, objective one, represented by traders and craftsmen working in the Aghabani industry. The number of interviews was chosen in accordance with the available time and resources based on previous social relations and personal acquaintances. The interviews took around an hour each and were preceded by an exploratory tour and a preliminary survey of the research community. We are grateful to Lana Mradni for her assistance in conducting and transcribing the interviews.

The research extracted a range of information on the responses of governmental and non-governmental institutions and the effectiveness of actions taken in the development process of the industry. The research found that the motivation of individuals and their adherence to the industry for its being part of their identity is indeed a form of cultural resistance. The craft's practitioners show flexibility in adapting to changes to ensure the preservation of their intangible cultural heritage and its transmission to future generations.

Introduction

In times of war, the search for identity and belonging is strengthened, and the factors, conditions and objects that establish such feelings become an urgent need to confront the daily scourges of death. Therefore, the importance of traditional crafts and heritage is emphasised at wartime, for they are of the factors and tools directly linked to the prewar collective identity and memory transmitted over generations to the present time, despite historical conflicts and changes, to remain as a proof of survival and the ability to survive.

The non-material, or intangible, cultural heritage is a target of the war for its being linked to the productive society. Its destiny is also linked to the ability of this society to absorb the disturbances and to reorganise itself to create an environment conducive to the transmission of this heritage's skills and experiences to future generations, thus safeguarding the non-material heritage from going extinct.

The war in Syria grew out of a popular uprising that began in 2011 and soon turned into a military conflict centred mostly in the countryside cities that were incubators of peaceful protests at first and of armed dissident groups later on, until the movement was dominated by factions called the armed opposition. Later on, government troops and allied forces besieged these cities for years, causing a sharp rise in prices and general hardships, including a deterioration in the healthcare sector and shortages of food, medical supplies and other daily necessities. One of these cities was Douma, a city of 200,000 people in the province of Damascus about 14 kilometres away from the centre of Damascus[1]. It was also one of the first districts to participate in the peaceful protests in 2011. A siege that began in 2012 ended with the return of government control of the city in April 2018 after a

---

[1] *Al Jazeera Encyclopedia*, "Douma: Ghouta's Bride Targeted by Blockade and Poisonous Gases" [electronic reference], available at https://bit.ly/2CSXqpN, published on April 7, 2018, accessed on January 24, 2019.

military campaign caused the death and displacement of many of its people[2].

Douma, along with the towns of Irbin, Saqba, Hamouriyah, Zamalka, Harasta and al-Marj area, in addition to dozens of villages and towns, form what is known as Damascus' eastern Ghouta region, which is connected geographically and directly to Damascus[3]. The women of Douma, also called the Doumanis, are known for their embroidery of the Aghabani while the men work in the crafts complementing this industry, namely the printing of designs on the fabric before embroidery and the maintenance of the embroidery machines[4]. This industry is historically associated with Damascus, which is a site for the distribution of products and the concentration of some of its craftsmen. As mentioned in many sources, these two cities are major centres for the manufacture and marketing of the Aghabani[5], which is an element of Syria's non-material cultural heritage (see Photo 1)[6].

---

[2] *Al Jazeera Encyclopedia*, "The Eastern Ghouta: A Syrian Region under Destruction" [electronic reference], available at https://goo.gl/Er45Bu, published on February 21, 2018, accessed on August 2, 2018.

[3] *Al Jazeera Encyclopedia*, "The Eastern Ghouta: A Syrian Region under Destruction".

[4] Talal Maallah, *The Non-Material Heritage: Skills Associated with Traditional Craftsmanship, Part I*, Damascus: Syrian Ministry of Culture and Syria Trust for Development, 2014, p. 62.

[5] See the official website of the Syria Trust for Development at https://bit.ly/2OOMsq4. See also Mounir Kayyal, *Levantine Achievements in Damascene Arts and Industries*, Damascus: Syrian General Authority for Books, Ministry of Culture, 2006–2007, p. 80.

[6] Talal Maallah, The Non-Material Heritage: Skills Associated with Traditional Craftsmanship, p. 60.

Photo 1: A form of Aghabani embroidery – the palm and rose design – from the "Threads of Hope – Khouyout Alamal – Aghabany" Facebook page[7]

The Aghabani, according to Mounir Kayyal's book, is "a cloth of cotton fabric with special characters, embroidered and embellished with silk or synthetic fibre threads. This fabric is one of the sources of pride for Damascene textiles"[8] and its craft was first known in Aleppo before it moved to Damascus[9]. Kayyal continues to describe the stages of the industry, which begin with preparing the yarns before weaving and dyeing them with the colours required by the "Mazayki"[10], stiffening

---

[7] From the "Threads of Hope – Khouyout Alamal – Aghbany" Facebook page at https://goo.gl/D2Mx1o.

[8] Mounir Kayyal, Levantine Achievements in Damascene Arts and Industries p. 80.

[9] Talal Maallah, The Non-Material Heritage: Skills Associated with Traditional Craftsmanship, p. 61.

[10] The "Mazayki" places iron poles and hangs the dyed silk yarns on them—the length of each yarn is not less than 21 arms—where he inspects them, repairs and connects any that were cut before sending them to the "Mulqi." *A Glossary of Ancient and Modern Professions until the Dawn of the Twenty-First Century, Section One: An Arabic-Arabic Dictionary* can be found at https://goo.gl/bVbVaw.

them with starch and sending them to the "Mulqi"[11]—literally, the thrower—who prepares the threads for the weaver (see Photo 2). Finally, the weaver starts weaving the cloth on the loom[12].

Photo 2: Silk Threads – From the "Threads of Hope – Khouyout Alamal – Aghabany" Facebook page[13]

Then comes the phase of printing the designs on the cloth, which are usually botanical drawings or Arabic motifs that are engraved onto special wooden blocks and are called al-Rash, al-Tals, al-Dhamma, Abu al-Hajb, waqf al-qaa' and other names depending on the drawings (see Photo 3)[14].

---

[11] al-Mulqi: An old Damascene profession where the Mulqi prepares the threads and delivers them to the weaver. A Glossary of Ancient and Modern Professions until the Dawn of the Twenty-First Century, Section One: An Arabic-Arabic Dictionary can be found at https://goo.gl/bVbVaw.

[12] Mounir Kayyal, Levantine Achievements in Damascene Arts and Industries, p. 80.

[13] From the "Threads of Hope – Khouyout Alamal – Aghabany" Facebook page at https://goo.gl/D2Mx1o.

[14] Mounir Kayyal, Levantine Achievements in Damascene Arts and Industries, p. 80.

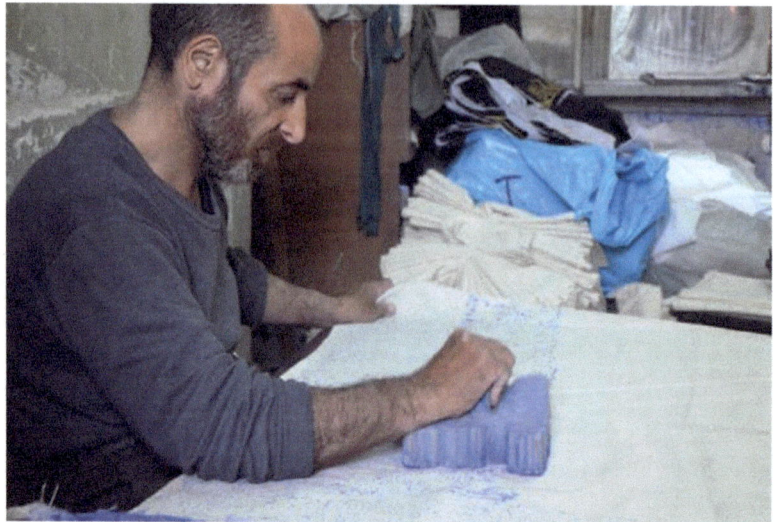

Photo 3: An Aghabani Printing Worker - Maher Al Mounes - Damascus – 2018

The printing worker prints the drawings several times on the piece of fabric before sending the cloth to the female workers (see Photo 4) to do the embroidery in a special style. Historically, the cloth was pulled onto a circular wooden hoop before the women started embroidering the drawing using silk threads and special needles. As the production process evolved, the wooden hoop and handwork were replaced by a machine similar to a household sewing machine[15].

The Aghabani industry goes through several stages: printing, embroidery, cleaning off the moulds' imprints, ironing the finished product and selling it in the markets of Damascus. It is an old Damascene industry. Most of the embroidery work has taken place in Douma since more than 100 years ago.

---

[15] Mounir Kayyal, Levantine Achievements in Damascene Arts and Industries, p. 80.

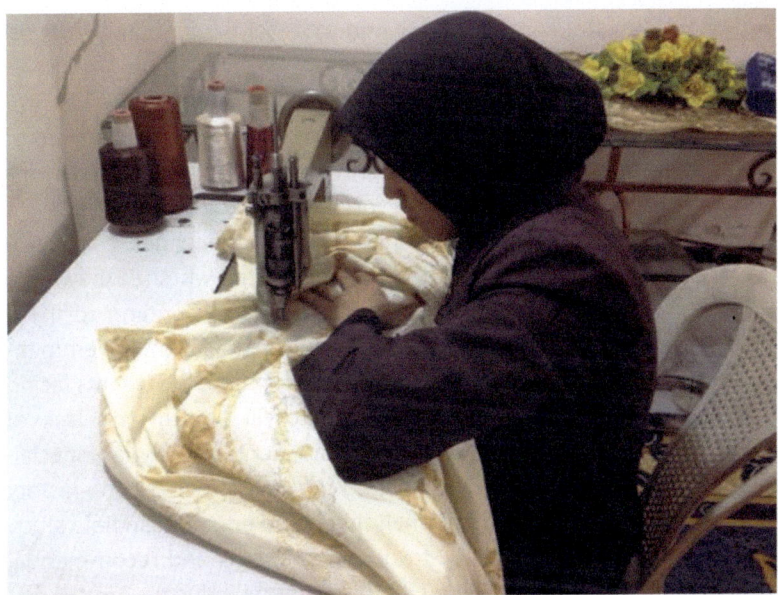

Photo 4: A female worker in the process of embroidery – from the "Threads of Hope – Khouyout Alamal – Aghbany" Facebook page[16]

The war conditions imposed on the traditional craft industries in general and the Aghabani in particular, as the city of Douma was besieged and experienced killing and displacement, caused a number of challenges, which directly affected the industrial society. Although the industry still exists in Damascus, production conditions have been changed because of a range of social, economic and logistical factors, the most important of which are related to the production process. These include the displacement of female workers, a lack of human resources and raw materials, and challenges of marketing the finished products.

This research addresses the questions whether male and female workers in the Aghabani industry are motivated to help find mechanisms and solutions that contribute to the preservation of the industry from extinction and whether the emerging challenges contributed to the enhancement of their heritage value, considering that the

---

[16] From the "Threads of Hope – Khouyout Alamal – Aghbany" Facebook page at https://goo.gl/D2Mx1o.

Aghabani industry is part of their non-material heritage and thus reflects their cultural identity.

This research attempts to contribute in highlighting the war impacts in killing and displacing the people of Douma, the destruction of civil life, impediments to economic growth, and the impact of the disruption of the transmission of this industry to future generations[17]. It also seeks to contribute to monitoring the challenges that have weighed on the actors in this traditional industry, whether they were displaced or based in Damascus, how they are responding to these challenges, and the solutions developed by them and by governmental and non-governmental stakeholders to overcome the threats facing the industry. Thus, the research seeks to fill part of the lack of knowledge related to the intangible cultural heritage in Syria, especially the Aghabani industry, by collecting information about the history and conditions of this industry, analysing it and engaging in field study in Damascus, as a case study, to obtain solutions and recommendations from the actors themselves, to maintain the industry.

The research relies on a qualitative methodology based on collecting information from the industry traders and craftsmen, an approach which, according to Natasha Mack[18], can gain a heavy and complex understanding of a particular social context or phenomenon through the ability to provide a qualitative description of people's experience within the research's studied problem. Qualitative research methods also provide information on the human aspect of an issue, including the behaviours, beliefs, opinions, emotions and relationships between individuals, which are often contradictory. Thus, this methodology will help in understanding the context of the Aghabani industry through interviews with people active in the craft before 2011 and at the time of completion of this research in 2018, contributing insights from the perspectives and experiences of members of the industry, in addition to gathering information to understand the human and social dimensions of the phenomenon that might not be available in some refer-

---

[17] Fayaz Ahamad and Effat Yasmin, "Impact of Turmoil on the Handicraft Sector of Jammu and Kashmir: An Economic Analysis", Department of Economics, University of Kashmir, Srinagar, India, *International NGO Journal*, Vol. 7(5), 2012, pp. 78–83.

[18] Natasha Mack et al., *Qualitative Research Methods: A Data Collector's Field Guide*, Research Triangle Park, North Carolina: Family Health International, 2005, p. 1.

ences, bridging the information gap and the scarcity of documentation for this industry.

This study is based on semi-structured in-depth interviews and the use of flexible open-ended questions (see Appendix 1), which were inspired by preliminary exploration and questioning of the commercial and productive entrepreneurs of the Aghabani industry at Damascus. The research sample is a non-random, objective one represented by male and female traders and artisans working in the industry. There were six in-depth interviews (see Table 1), based on previous social relationships. Two traders have been contacted, both of whom have businesses in manufacturing and marketing the Aghabani, and a former female technical director who used to work in embroidery in a project that manufactures and sells the Aghabani. Through these people, communication was made with other craftsmen such as a printing worker, a repairman of embroidery machines and a person who works in embroidery. Each interview lasted for approximately one hour, preceded by an exploratory tour and a preliminary survey of the research community. It is worth mentioning that there is inconsistency in some of the information provided by members of the research sample in relation to the history and origin for the industry and activity before and after 2011. It should also be noted that we will not use the full names of the people interviewed, but we will identify them only by nicknames or professional qualities, because of the sensitivity of the information they provided and their wish not to disclose their full names.

Table 1: Information on the sample members.

Job Title	Age Group	Duration of work in this craft	Number of generations working in the industry	Do your sons work in the industry?
Abu Bassam, a merchant (nickname)	50–65	More than 40 years	3 generations	Only him; his sons tried to work with him but did not continue
Abu Samer, a merchant (nickname)	40–50	More than 30 years	3 generations	He works along with his brothers and cousins

Abu Abdo, a printing worker (nickname)	40–50	More than 20 years	1 generation	Only him
Abu Muhammad, an embroidery worker (nickname)	30–40	More than 20 years	1 generation	Only him
Abu Saeed, a machine repairman (nickname)	50–65	More than 40 years	3 generations – his father opened a shop for repairing embroidery machines in 1936 in Douma, encouraged by his family	Only him – however, his son is planning to work with him after completing his university studies
The Doumaniya ("the Doumani woman," a nickname), a technical director and former embroiderer	50–65	More than 30 years	3 generations	Only her; she has no children

We reviewed a number of bibliographies and books to collect documentary information about the industry's history and place of origin, such as Mounir Kayyal's *Levantine Achievements in Damascene Arts and Industries*[19], Najat Qassab Hassan's *Damascene Discourse*[20], and Mohammed Fayyad al-Fayyad and Majed Hashem Al Hammoud's *Traditional Crafts in Syria*[21]. We also consulted Talal Maallah's book on *The Non-Material Heritage*[22], which is considered one of the few sources that document the current state of the industry. We have also relied on academic essays that explain the theories of non-material heritage, its relevance to the identity of the societies producing it and the mechanisms of its preservation, such as the essays of Graciela G. Singer[23] and

---

[19] Mounir Kayyal, Levantine Achievements in Damascene Arts and Industries, 2006–2007.

[20] Najat Qassab Hassan, *Damascene Discourse: Memoirs 1, 1884–1983,* Tlass Publishing House, Damascus, 1988.

[21] Mohammed Fayyad al-Fayyad and Majed Hashim Hammoud, *Traditional Crafts in Syria*, first edition, translated by Majd Hamoud, Damascus: General Union of Craftsmen, Office of Culture and Media, 2011.

[22] Talal Maallah, The Non-Material Heritage: Skills Associated with Traditional Craftsmanship, pp. 61–62.

[23] Graciela G. Singer is a Near East historian who focuses on issues related to the history and archaeology of Bronze Age Egypt and the Eastern Mediterranean.

Federico Lenzerini[24]; and academic books and articles focusing on theories of cultural resistance and resilience using art and handicrafts in times of conflict, such as those of Patricia Leavy[25] and Dietrich Heissenbüttel[26]. In addition, some online news articles helped in investigating the roles of governmental and non-governmental actors at different periods, analysing their effectiveness, and identifying and categorising the reactions of workers according to the aforementioned theories of cultural resistance and flexibility.

The research begins by reviewing the theoretical reference by defining a set of concepts related to cultural resistance and cultural identity in times of war, the importance of cultural resistance through handicrafts and their impact on the preservation of non-material heritage, as well as discussing the lack of knowledge about the Aghabani industry, which included information on the industry itself prior to 2011.

The first part of the research deals with the historical context of the Aghabani industry since its beginning, using the information collected from the sample and from books and references to understand the origin and development of the industry. The second part is an analysis of the response mechanisms of official and non-official institutions, the craftsmen's union, UNESCO and industry workers through the analysis of interviews with sample members and reviewing the online articles. Finally, the research proposes a set of recommendations aimed at supporting and empowering the workers in this industry and developing and preserving their knowledge and skills as a non-material cultural heritage.

---

[24] Federico Lenzerini is a professor of international law and European Union law at the University of Siena's Faculty of Law. He is a consultant to UNESCO's Department of Cultural Heritage Protection and is the legal counsel to the Italian Ministry of Foreign Affairs for the conduct of international negotiations concerning the protection of cultural property.

[25] Patricia Leavy is an independent scholar, author and public speaker.

[26] Dietrich Heissenbüttel has studied literature and art history. Since 1996, he has been preparing radio programmes on jazz and improvised music in France. He also contributes to several newspapers and magazines, specialising in globalization, contemporary art, New Age music, architecture and transport policy. He has also taught at the Institute of Art History at the University of Stuttgart since 2008.

One of the difficulties we faced was identifying the research community in Damascus, for it was impossible to conduct any interviews inside Douma at the time of study[27] because of the security conditions, the destruction of the city and the displacement of the majority of its population internally, primarily to Damascus or Idlib[28]. Other difficulties included the paucity of research dealing with the Aghabani industry during the war that we managed to access, and the small number of references that focused on the Aghabani. References were limited to undocumented personal narratives or historical narratives of handicrafts in general, which briefly describe the Aghabani industry in Syria. Besides, there is a lack of official articles that focus on the Aghabani industry specifically, as many of them deal with all handicrafts, as well as a lack of references and studies dealing with the topics of cultural resistance and flexibility. The importance of this research lies in its effort to fill the knowledge gap related to the history of the Aghabani industry and to compare its status before and after 2011 in terms of change in the dynamics of production, the analysis of the motives of workers to continue the Aghabani industry, the possibility of classifying these motives as acts of cultural resistance, and analysis of stakeholders' responses and their impact on the industry based on the theories of safeguarding the intangible cultural heritage.

---

[27] The research was conducted between February and May 2018.
[28] *Arab48*, "A Mass Exodus from Douma and Thousands of People Leaving Ghouta for Idlib" [electronic resource], on the *Arab48* news website, available at https://goo.gl/RTF9tL, published on March 26, 2018, accessed on May 23, 2018.

Definitions and Terminologies

In this research, we will adopt a set of conventional terminologies used in the field of the Aghabani industry as they are used on the ground. Given that these terms are sometimes closer to the colloquial language than the classical one, it was necessary to define and describe these terms before proceeding into the chapter on the study itself.

The printing worker: This is the person who uses wooden moulds to stamp the botanic shapes and other decorations on the cloth before the process of embroidery.

The female workers ("Shaghalat"): These are the women who work in embroidery. Women from Douma are well-known in this profession[29].

The Craftsman: This is a person who works in the production of materials or who provides service materials depending primarily on his personal effort and professional experience.

The Artisan: This is the employee who is proficient in his craft, but still needs time to fully master it. Thus, he works with a teacher ("mu'allim") after agreeing on the work duration and the wage[30].

The Doumaniya: This term is used among the workers of the Aghabani industry to describe a woman from Douma who works in this industry. The people of Douma are called the Dwamna, or Doumanis, in Syria. "The Doumani" refers to a man from Douma and "the Doumaniya" stands for a woman from Douma[31].

The embroidery machine: This is the specialised sewing machine used by female workers to embroider the cloth. It is powered either manually or electrically[32].

---

[29] Mounir Kayyal, Levantine Achievements in Damascene Arts and Industries, pp. 80, 82.

[30] Mustafa Dandashli, "Dr. Abdul-Karim Rafik: The Manifestations of Craftsmanship Organization in the Levant in the Ottoman Era" [electronic reference], the Cultural Center for Research and Documentation in Sidon, at https://bit.ly/2v4gq0i. Published on April 1, 1981, accessed on September 15, 2018.

[31] Sabreen al-Sa'o, "The City of Douma in Syria" [electronic reference], available at https://goo.gl/sgYsTM, published on April 4, 2017, accessed on August 12, 2018.

[32] Lana Mradni (the author), an interview with the Doumaniya (a nickname) about the Aghabani industry in Damascus, (n.p.), Damascus, 4/28/2018.

The lathe: This is a machine used in the manufacture of metal spare parts used in the maintenance of industrial machines.

Coordinating Councils: These are the regulatory bodies that emerged in Syria's regions that protested against the regime in 2011. Their names were associated with the names of the regions, where a small group of young people responsible for organising and mobilising demonstrations, coordinating their work, slogans, and securing logistical and media support[33].

Consumer Protection Committee: A body affiliated to the Ministry of Interior whose role is to determine and control prices in the market according to the cost in order to protect the interests of the consumer[34].

---

[33] al-Asaad Omar, "Creative Forms in the Syrian Revolution: The Coordinating Councils as a Model" [electronic reference], on *Babelmed*, available at https://goo.gl/jyYqEz. Published on June 20, 2012, accessed on January 25, 2019.

[34] Syrian Arab Republic: Syrian Prime Ministry, Law No. 14 of 2015 on Internal Trade and Consumer Protection [electronic reference], from the website of the Prime Ministry of the Syrian Arab Republic, available at https://goo.gl/u6B4uq, accessed on April 3, 2019.

Literature Review

In this section of the research, we will present the references and studies that we have adopted in theoretical terms, whether those documenting the Aghabani industry or those discussing the theories that helped explain some of the phenomena and concepts related to the culture of resistance and non-material heritage and the extent to which the assets of these theories correspond to the Syrian situation in general and the Aghabani industry in particular.

Among the references on which the research was based is a book by the historian Mounir Kayyal titled *Levantine Achievements in Damascene Arts and Industries*[35], which includes a historical account of the traditional industries in terms of technical and manufacturing processes as well as information on their origin and history. In his book, Kayyal dedicated a section to the Aghabani industry in a historical descriptive narrative style on the types of fabrics and embroidery drawings, in addition to documenting the steps of manufacturing, from preparing the thread to the process of printing on cloth and sending them to female workers for embroidery. Kayyal refers to the fame of the women of Douma, located in Damascus' countryside, in the embroidery process[36]. In an interview with eSyria, Kayyal said that "the environment and the people are his first sources of information"[37], adding that he relied on the descriptive method in compliance with the advice of the historian Muhammad Dahman. He also relied on some manuscripts and sources. Despite the writer's motivation to document the Damascene industries, his work is not based on statistical information. For example, the author mentions the figures without documenting the source of such information. He said that 90 percent of the Aghabani's

---

[35] Mounir Kayyal, Levantine Achievements in Damascene Arts and Industries, 2006–2007.
[36] Mounir Kayyal, Levantine Achievements in Damascene Arts and Industries, pp. 80–82.
[37] Ismail al-Najem, "Mounir Kayyal: A Documented Encyclopedia as a Tribute to the City: 20 Books Telling the Details of the Levant" [electronic reference], on eSyria.sy, available at https://goo.gl/HVGcKh, published on November 20, 2014, accessed on May 23, 2018.

production is exported to foreign markets[38] without mentioning a reference to support that figure, which makes it impossible for us to verify the accuracy of this figure and its significance. In spite of this, some of his writings were references in official journals and other references[39], and his book *Levantine Achievements in Damascene Arts and Industries* was considered an important documentary field study sponsored by the Syrian Arab Republic's Ministry of Culture[40].

One of the references that documented the Syrian cultural heritage in writings and photographs is Johannes Kalter's *The Arts and Crafts of Syria*[41]. However, this book's discussion of the Aghabani is limited to a historical narrative of fashion and the use of the Aghabani to highlight social status during the Ottoman era.

The research has also relied on Talal Maallah's book *The Non-Material Heritage: Skills Associated with Traditional Craftsmanship*[42], which classifies the Aghabani industry as one of the 100 elements of Syria's non-material heritage. The industry originated in Aleppo before it moved to Damascus. The process of embroidering the Aghabani is done by women in some rural areas such as the city of Douma, while men work on printing the designs on the cloth. The book also states that the Aghabani industry, in all its stages, is transmitted within the same family from mothers to daughters and from fathers to sons[43], and goes on to mention the dangers threatening the craft: such as the decline of local production of raw materials, the decline of interest among young people to learn the industry or trades related to it, the lack of systematic documentation of the work's stages, which has led to a decline in the number of people familiar with the contents of each

---

[38] Mounir Kayyal, *Damascene Arts and Industries*, Syrian Ministry of Culture, Damascus, 1985, p. 113.

[39] Adra, Hisham, "Syria's Handicrafts and Traditional Knowledge Are at Risk of Loss" [electronic reference], on *Asharq al-Awsat* website, available at https://goo.gl/F95Lgj. Published on February 3, 2015, accessed on June 28, 2018.

[40] Mounir Kayyal, Levantine Achievements in Damascene Arts and Industries.

[41] Johannes Kalter et al., *The Arts and Crafts of Syria*, London: Thames and Hudson, 1992, pp. 233, 499.

[42] Talal Maallah, The Non-Material Heritage: Skills Associated with Traditional Craftsmanship, pp. 61–62.

[43] Talal Maallah, The Non-Material Heritage: Skills Associated with Traditional Craftsmanship, p. 63.

stage, and finally the decline in the number of professionals in the manufacture of the wooden moulds used in printing.

Despite its importance, Maallah's book, published in 2014, does not give a clear image of the contemporary dangers and threats to traditional crafts resulting from the deterioration in the security situation in the cities of Damascus and Douma. It does, however, discuss the reasons and circumstances threatening the industry before the war.

Our study also relied on the UNESCO Convention for the Safeguarding of the Intangible Cultural Heritage[44], which was ratified by the Syrian Government, making it binding to its terms. This globally ratified agreement emphasises the importance of safeguarding the intangible cultural heritage as a human right and considers any harm affecting the culture and non-material heritage of any people as a damage to the cultural heritage of humanity as a whole[45]. This research seeks to shed light on the role of UNESCO in the conflict in Syria through a series of statements and initiatives undertaken to mitigate these effects and to oblige the conflict parties to be committed to the international conventions on the protection and preservation of cultural heritage.

The research also focuses on the theories that emphasise that the intangible cultural heritage is not related to the product itself—in this case, the finished embroidered Aghabani textile—but to the intangible factors associated with the society in which the process of manufacturing the product takes place, such as the knowledge and skills transmitted from one generation to another and the resulting social and economic values that are important for the development of socie-

---

[44] UNESCO (United Nations Educational, Scientific and Cultural Organization), "Convention for the Safeguarding of the Intangible Cultural Heritage, Article 2: Definitions", adopted in Paris on October 17, 2003, available on the UNESCO Intangible Cultural Heritage website at https://bit.ly/2OKUobK.

[45] University of Minnesota, "The Hague Convention for the Protection of Cultural Property in the Event of Armed Conflict" [electronic reference], University of Minnesota website, available at https://bit.ly/2GnnMRZ, accessed on May 23, 2018. (For an English-language version of this document, see UNESCO, "The 1954 Hague Convention for the Protection of Cultural Property in the Event of Armed Conflict" [electronic reference], on UNESCO's Armed Conflict and Heritage website, available at http://bit.ly/2xiSUxK.)

ty[46]. The founding factors of this heritage are also represented in its self-identification as a fundamental element of the cultural identity of its creators and holders, on the one hand, and, on the other, in how this pushes them to continue to re-create it in response to the historical and social development of societies and groups to which they belong and their attachment to its cultural identity, reflecting its originality of this heritage and confirming its inherent relationship to human rights[47]. This is what constitutes the essence of groups and their distinction from others, which gives diversity to different cultures and constitutes a basis for the exchange of knowledge, development and creativity. This is vital to humanity and a cause of conservation called for by the UNESCO conventions[48]. According to Graciela Singer, the intangible cultural heritage takes its form only by the recognition of the groups that produce and transmit it continuously, and no one else, not even the state, can decide that on their behalf. Therefore, they should be appreciated as the protectors of the collective memory[49], defined by Maurice Halbwachs[50] as "the work of individuals' minds in society, which is defined and organised by a series of social arrangements"[51], and which is a contribution, according to Chiara De Cesari and Ann Rigney[52], in shaping the identity of individuals by belonging to a society and to a common memory[53]. This memory includes the common past, the memory of the place and the events associated with it, and the so-called material and non-material cultural heritage that

---

[46] Graciela G. Singer, "The Importance of the Intangible Cultural Heritage" (electronic reference), on the blog *Terrae Antiqvae*, available at http://bit.ly/2XH736y, published on January 5, 2011.
[47] Federico Lenzerini, "Intangible Cultural Heritage: The Living Culture of Peoples", *European Journal of International Law*, Vol. 22, No. 1, 2011, pp. 101–110.
[48] UNESCO, "Convention for the Safeguarding of the Intangible Cultural Heritage".
[49] Graciela Singer, "The Importance of the Intangible Cultural Heritage".
[50] Maurice Halbwachs was a French philosopher and sociologist known for developing the concept of collective memory.
[51] Maurice Halbwachs, *On Collective Memory*, translated and edited by Lewis A. Coser, Chicago, Illinois: University of Chicago Press, 1992, p.38.
[52] Chiara De Cesari is an anthropologist and an assistant professor in European studies and cultural studies at the University of Amsterdam. Ann Rigney is the head of comparative literature department at the University of Utrecht.
[53] Chiara De Cesari and Ann Rigney, eds., Transnational *Memory*: Circulation, Articulation, Scales, Walter de Gruyter GmbH, Berlin/Boston, 2014, p. 2.

reflects their cultural advantages. Thus, Lenzerini emphasises the need for the state to play its role in ensuring the continuity of these individuals and societies and to ensure their freedom to circulate and manage their intangible cultural heritage as an integral part of their rights[54].

The research attempts to determine the form of response of the actors in the Aghabani industry in Damascus and the changes in their motivation and awareness in order to explore whether they fall under the framework of cultural resistance. We look initially at the concepts of resistance in general, which are, according to Jocelyn A. Hollander[55] and Rachel L. Einwohner, "the collective action representing a direct challenge to power systems"[56]. In addition, cultural resistance in particular is a mechanism through which a community can formulate other solutions to problems and dilemmas, being the repository of imagination necessary for change[57], according to Simon Sweeney[58].

In the search for articles and studies on cultural resistance, we found that the term "cultural resilience," as defined by Stephanie J. Rotarangi and Janet Stephenson[59], is widely used in studies that examine the ability of individuals and societies to harness hope and optimism in the capacities generated by a community or cultural system to be able to absorb turbulence and reorganise during the process of change to maintain primarily the essential elements of organization and identity[60] that crystallise the particular characteristics of this community. In addition, the research adopted the theories that emphasise the symbolic use of handicrafts and textiles, which are, according to

---

[54] Federico Lenzerini, "Intangible Cultural Heritage: The Living Culture of Peoples", pp.101–103
[55] Jocelyn A. Hollander is a professor of sociology at the University of Oregon.
[56] Jocelyn A. Hollander and Rachel L. Einwohner, "Conceptualizing Resistance", *Sociological Forum*, Vol. 19, No. 4, December 1, 2004, p. 538.
[57] Simon Sweeney, "Review: Reconstructing Spain: Cultural Heritage and Memory after Civil War", *International Journal of Heritage Studies*, Vol. 17, No. 6, November 2011, pp. 629–631.
[58] Simon Sweeney is an English author and lecturer in international political economy and business at York University.
[59] Janet Stephenson is director of the Centre for Sustainability at the University of Otago and director of the National Energy Research Institute of New Zealand.
[60] Rotarangi, Stephanie J., and Janet Stephenson, "Resilience Pivots: Stability and Identity in a Social-Ecological-Cultural System", *Ecology and Society*, Vol. 19, No. 1 (March 2014), available at https://goo.gl/uWek8P.

Clive Edwards, kinds of arts and forms of social and political resistance supported by the theory of Patricia Leavy that considers these types of arts as a different narration of war history[61]. We also adopted Dietrich Heissenbüttel's theory that sees the art of handicrafts a way to present personal narratives, in times of conflict and war, highlighting a more humane aspect of the war in the form of resisting death, the obliteration of identities and the erosion of heritage[62].

Finally, it should be noted that the research has drawn on a number of articles published on websites like *Al-Thawra Online* and *Tishreen*, which are official newspapers that reflect some criticism to the Syrian government's internal policies and the measures taken in response to the war to preserve the intangible cultural heritage or to continue securing some elements of the Aghabani industry, like the cloth. In addition, we have obtained information from news articles on Arab and foreign websites such as *Al-Akhbar, Al Jazeera, Al-Jumhuriya*, and the BBC, among others.

---

[61] Patricia Leavy, *Method Meets Art: Arts-Based Research Practice*, New York, NY: Guilford Publications, 2015, p. 216.

[62] Dietrich Heissenbüttel, "Art in Conflict: Interventions in War and Crisis", an essay published in the context of the Global Activism exhibition in Karlsruhe, Germany, 2014, p. 4.

Part One –
The Historical Context of the Aghabani Industry and the Changing Dynamics of Its Work After 2011

In this section, the research identifies some of the features regarding the origin of the Aghabani industry, based on references and information gathered from interviews with members of the study sample. Then, it summarises the industry's working conditions before 2011 and the series of changes that resulted from the war.

1. A Brief History of the Aghabani's Origin

The interviewees agreed that the Aghabani industry was concentrated in the city of Douma. However, their views varied on details of the industry's move to other nearby cities. While the merchants Abu Bassam and Abu Samer and the machine repairman Abu Saeed believed that the work moved to the city of Harasta through marriage, the Doumaniya stressed her belief that the movement of craft was confined to the villages surrounding Douma, such as Mesraba, without reaching neighbouring cities like Harasta, with a possibility of its transfer to the city of Jisreen (see the map of the Damascus countryside, Figure 1) [63]. The Doumaniya also stressed that the industry was part of the traditions and customs of the city's lower classes of peasants, while the notable families refused to let their daughters work in the Aghabani, deeming it something shameful, that is to say, devaluing these girls. This may be because it indicates the poverty of the family and the inability to provide the needs of the girls, forcing them to work. It is also a source of noise, due to the loud voice of the embroidery machines. The marriage customs in the city dictate that the bride's family should provide their daughter with her household needs upon getting married,

---

[63] See a map of the Damascus countryside to find the locations of the listed cities, available on Google Maps at https://goo.gl/iWBtN2.

including some furniture, kitchen utensils, and the bride's clothes[64], which requires a large sum of money. Hence, the mothers in poor families begin to teach their daughters to work in the Aghabani from a young age, starting around 8 years old, so the girls can save money to buy their needs when they get married. This led, according to the Doumaniya, to the industry's becoming a "part of the region's culture and heritage, where it is rare to find a house that does not contain an Aghabani machine, and all women almost always mastered this style of embroidery"[65]. This is consistent with other references talking of the fame of the city's women in the Aghabani embroidery.

Figure 1

---

[64] Ahmad al-Allaf, *Damascus at the Beginning of the Twentieth Century*, translated by Ali Naisse, Damascus: Syrian Ministry of Culture, 1976, pp. 118–119.

[65] Lana Mradni (the author), "On the Aghabani industry in Damascus", an interview with the Doumaniya (a nickname), (n.p.), Damascus.

Some of the answers regarding Aleppo's relationship with the industry were compatible, as it is believed that the designs used in the Aghabani embroidery in Syria originated in Aleppo, where the first wooden printing moulds came from. However, the origins of these designs are believed to be from East Asia, and from India in particular. This corresponds with information in Mounir Kayyal's book in his remarks on the Aghabani, where he writes, "First, it was a tradition of the Indian belt. It was found in Aleppo, where there were 717 factories"[66]. But the same statement suggests that the Indian embroidery itself was produced in Aleppo's factories, and this is contrary to other sources that say embroidery in general was one of the familial occupations in which women have been working since AD 1750, and that the signs of modernization of the local production system in Aleppo—that is, new equipment and technologies developed in the West such as steam engines, individual weaving machines, industrial dyeing, printing, synthetic yarn, etc.—were not introduced until the beginning of the nineteenth century. From the aforementioned, we conclude that Mounir Kayyal's reference did not provide reliable information regarding the relationship between the embroidery of Aghabani in Aleppo and its origins, for the embroidery profession is older than the presence of the plants in Aleppo, although it has somehow influenced the work of embroidery in Damascus. That occurred certainly before the nineteenth century, according to the information we got from the interviewees, such as the merchant Abu Bassam, who confirmed some traders' attempt to transfer the Aghabani industry to Aleppo. "However, they did not succeed because of the poor embroidery work, although the silk and printed fabrics were transported from Damascus," he said[67].

Upon asking members of the study sample about the origins of the industry in Douma, the merchant Abu Samer claimed that it was founded thanks to his grandfather, who brought 30 French embroidery machines in the early 20th century, specifically in 1910, and distributed them to peasant women in his agricultural lands in Douma, so they could work on them in the wintertime. As the women got used to

---

[66] Mounir Kayyal, Levantine Achievements in Damascene Arts and Industries, p. 80.
[67] Lana Mradni (the author), "On the Aghabani Industry in Damascus", an interview with the merchant Abu Bassam (a nickname) (n.p.), Damascus, on March 3, 2018.

the work, he brought in more machines. Thus the industry spread and became an important part of the city's inherited cultural pattern, according to Abu Samer[68].

The Doumaniya's information contradicted that account. For her, it was likely that the industry entered Douma before 1910. She confirmed that many members of her family were proficient in the Aghabani craft. "The industry is an old one since the time of the Ottomans," she said, "since my grandmother taught me how to fold clothes' packages[69] she embroidered herself at the time of the Seferberlik[70] when she was working in the industry before her marriage. She was born before 1900"[71]. The Doumaniya went on to confirm that the embroidery of the Aghabani by the city's women was a source of wealth for some families whose men traded in embroideries. "My grandfather's sister did not get married and brought up her sisters," she said. "She worked in the Aghabani all night long and gave her sisters what she had produced to sell, so my grandfather became a feudal lord in Douma and was always rich in the time of the Seferberlik"[72]. Counting the generations to which the women mentioned in the Doumaniya's narration belong, the Aghabani industry would go back to four generations, or to 1827.

Similar discrepancies appear in the theoretical and historical references, as the research did not reach any definitive information about the industry's history and place of origin. Instead, it found a collection of information documenting the social life in Syria, and in Damascus in particular, where the Aghabani enters into daily details such as fashion, social status and craftsmanship organization. For example, a book by Johannes Kalter and others, titled *The Arts and Crafts of Syria*, documents the use of the Aghabani in men's fashion in the Ottoman Em-

---

[68] Lana Mradni (the author), "On the Aghabani Industry in Damascus", an interview with the merchant Abu Samer (a nickname) (n.p.), Damascus, on April 21, 2018.
[69] Buqaj: a package of clothes.
[70] The Seferberlik: The conscription practice of the Ottoman Empire during the World War I period (1914–1918), when many men in the Levant were compelled to join the Ottoman army in forced labour and fighting in its ranks.
[71] Lana Mradni (the author), "On the Aghabani Industry in Damascus", an interview with the Doumaniya (a nickname)
[72] Lana Mradni (the author), "On the Aghabani Industry in Damascus", an interview with the Doumaniya (a nickname).

pire, as well as brides' dresses. It also provides pictures of fashion in Syria dating back to 1870 and 1880, although the pictures of fashions with the Aghabani embroidery are not documented by any date[73] (see Photos 5 and 6).

Photos 5-6: Fashion with embroidery similar to that of the Aghabani. The reference did not clarify the type of the embroidery or its date[74].

The Aghabani is also mentioned in a book titled *Damascene Discourse*, a memoir by Najat Qassab Hassan[75] covering the period from 1884 to 1983. In his comments about the old organizational scheme of the artisanal trades, Qassab Hassan says: "The artisan cannot work independently until the sheikh ("chief") of craftsmen will authorise him to do so after an examination in which the sheikh assesses the artisan's proficiency and ability to work independently. After that, the artisan would put the Aghabani turban on his head; an evidence of the completion of his ability, that is, he holds a visible testimony on his head presented to him in a ceremony"[76]. The writer then continues with the narrative of his father, who was forced to take responsibility for his family after the death of his father, a death that was shortly followed

---

[73] Johannes Kalter et al., *The Arts and Crafts of Syria*, pp. 233, 499.
[74] Johannes Kalter et al., *The Arts and Crafts of Syria*, pp. 233, 499.
[75] Najat Qassab Hassan, Damascene Discourse: Memoirs 1, 1884–1983, p. 260.
[76] Najat Qassab Hassan, Damascene Discourse: Memoirs 1, 1884–1983, p. 260.

by the Seferberlik. This account provides further evidence of the existence of a craftsmanship system in the Ottoman era in which the craft was transmitted within a hierarchy that ascends from the employee to the artisan, the teacher and the sheikh of the craft. The origin of this system is a collection of laws brought in by Sultan Suleiman the Magnificent during his long reign (1520–1566). These laws were derived from the European occupational systems, which then developed internally in the sixteenth century until the Ottoman system gained its general organization and craftsmanship's hierarchy[77].

Thus, the precise date of the Aghabani's origin is a complex matter that requires a thorough historical research. However, we can confirm that the industry emerged in the eighteenth century and not the beginning of the twentieth century, and what is important here is to try to understand the origins of this industry and its relation to the city of Damascus and its countryside in the near period before 2011. This is what we will review in the coming paragraphs.

2. The Aghabani before 2011

In the past, the Aghabani industry prospered in the city of Douma, where the machines' sound was heard in the neighbourhoods upon passing through. The work was coordinated between the working women and the merchant by a female supervisor who distributed the work to the women, each according to her work skill, giving small pieces to trainees and large ones to the more skilful women. Upon coming from Damascus, the merchant would find a city full of energy, as the female workers would come to one of the houses to give what each of them had embroidered to the supervisor. The latter would, in turn, watch and assess the wages paid according to the number of pieces and return to the worker any piece that needed to be repaired.

However, the number of female workers in Douma, according to the Doumaniya, has been decreasing since before 2011. The sound of machines could still be heard while strolling the alleys of Douma, yet

---

[77] Nasaar Abdul-Azim, "Craftsmanship Organization and Local Industries in the Ottoman Era" [electronic reference], *Journal of the Kufa Center for Studies*, Issue 43, available at https://bit.ly/2HYCiCG, published in 2016, accessed on April 4, 2019, pp. 195–240.

the work in this craft had become rare[78]. The decline in the Aghabani work in the past twenty years can be attributed to several reasons, some of which can be summarised as follows:

1) A trend among female workers to move away from the industry and start working in other handicrafts.
2) The transformation of the residential system from the traditional Arab houses to vertical buildings, which caused neighbours to be disturbed by the loud sound of machines.
3) The increased rate of education for girls, who previously were dedicating their time since an early age to work in embroidery.
4) Women are now engaged in other handicrafts that compete with the Aghabani work, such as beading, leather handicrafts and making coloured canes. In these crafts, the women sit with their neighbours, working together and engaging in chatting sessions.

The handicrafts in general have faced several challenges, including: the opening up of markets for imported industrial products[79] whose lower prices enabled them to compete with local handicrafts; high production costs; difficulties in marketing and the high cost and number of fees and taxes compared to the low income of craftsmen. These and other factors led to a decline in trade and have driven many craftsmen to work in other businesses, since the government taxes and fees did not take into account the funding difficulties, the cost of importing of raw materials, the competition with the machine-made products and the existence of cheaper alternative products due to the efforts of some foreign states to steal certain traditional crafts by reproducing and selling them in domestic and international markets in a way defaming the reputation of the local industry. In addition, some of

---

[78] Lana Mradni (the author), "On the Aghabani Industry in Damascus", an interview with the Doumaniya (a nickname).

[79] Bashar al-Hajali et al., "Traditional Crafts and Industries: A Homeland's Memory. How Can We Restore Its Spark?" [electronic reference], *Al-Thawra Online*, available at https://goo.gl/7nPGqx, published on December 22, 2012, accessed on September 7, 2018.

the points on the role of the craftsmen's union in responding to the risks faced by craftsmen in general will be clarified in Section 2.

The transformation of the residential system from traditional Arab houses to vertical buildings also created problems, because neighbours were disturbed by the sound of the Aghabani embroidery machine and complained against the female workers. However, work at Arab-style houses continues.

Upon searching for sources that reflect the reality of the urban development of Douma before 2011, we found a collection of articles in *Tishreen*[80], a Syrian state newspaper, the Syria News[81] website and the Rational and Guided Municipality Forum[82] that discuss the issue of urban planning in the areas around Damascus, including the Ghouta region, where Douma is located in its eastern part. These articles agreed on the problem of random urban growth, which means that "a group of residential areas, including buildings constructed without a license, whether in its inception, is contrary to the urban character of the areas on which it was established, defined in the general organizational chart of Damascus in 1967 and earlier as agricultural lands, green areas or industrial zones"[83]. In Ghouta (the traditionally green

---

[80] "On World Environment Day, the Urban Expansion of Damascus' Ghouta Claims 4,000 Hectares of Agricultural Land" [electronic reference], *Tishreen* newspaper, available at https://goo.gl/ov73MN, published on June 7, 2006, accessed on May 10, 2018.

[81] Kholouf, Hassan, "The Surroundings of Damascus and Urban Havoc", [electronic reference], Syria News website, available at https://goo.gl/V2EV9L, published on January 6, 2011, accessed on September 7, 2018.

[82] Safdie, Hossam, "The Expansion of Damascus between the Dreams of Strategic Vision, the Tragedy of the Nearby Countryside and the Reality of a Japanese Study" [electronic reference], from the Rational and Guided Municipality Forum, available at https://goo.gl/g8v16V, published on February 27, 2011, accessed on July 15, 2018.

[83] Adnan Masri, "The Impact of Urbanization in Damascus on Passenger Transport Movement: A Ph.D. Thesis in Economic Geography", Damascus University, Faculty of Arts and Humanities, Geography Department, 2014–2015, available at https://goo.gl/M2FPwc.

area)[84], a serious phenomenon emerged, where random housing engulfed 4,000 hectares of the region between 1965 and 1994[85].

These articles accused the administrative and technical staff of the municipalities and the governorate of Damascus of direct and indirect responsibility for the diffusion of this phenomenon. In his article titled "The Surroundings of Damascus and Urban Havoc", Hassan Khalouf wrote, "Exceptions signed by former governors ... in return for large sums of bribes, has contributed to the destruction of Damascus' Ghouta"[86].

Other independent sources addressed the same issue deeming "the phenomenon of random housing a social bribe when the Assad regime lacked the material means to finance housing, as it was the case in the time of the powerful Baath"[87]. This made the regime of former President Hafez al-Assad administratively lenient to disregard random construction with its promises to improve the conditions of poor housing services like electricity and water. Such promises were still not accomplished at the outbreak of the events of 2011. Other sources said that the Governor of Damascus and his staff had benefited financially from overlooking or granting construction licenses, which allowed urbanization to spread at the expense of agricultural land in Ghouta, as well as "the extensive acquisition of lands in Ghouta by successive governments to build government buildings, housing, facilities and services without compensation for the owners of the expropriated lands"[88].

We noted that most of these references, whether they were articles published by official governmental bodies or independent ones, agreed that the phenomenon of random urbanization has spread dur-

---

[84] Hossam Safdie, "The Expansion of Damascus between the Dreams of Strategic Vision, the Tragedy of the Nearby Countryside, and the Reality of a Japanese Study".
[85] "On World Environment Day, the Urban Expansion of Damascus' Ghouta Claims 4,000 Hectares of Agricultural Land".
[86] Hassan Kholouf, "The Surroundings of Damascus and Urban Havoc".
[87] Fabrice Balanche, "A City under the Control of the Baath: Damascus, the Syrian Capital" [electronic reference], *Al-Jumhuriya* website, available at https://goo.gl/RWpWf9, published on June 14, 2012, accessed on May 27, 2018.
[88] Samer Kakerli, "With Figures, Orient Net Investigates How Assad Destroyed Damascus' Ghouta" [electronic reference], Orient Net, available at https://goo.gl/AQuqtj, published on September 8, 2014, accessed on June 6, 2018.

ing the rule of the Assad's son, President Bashar al-Assad, because of administrative corruption in the heart of government institutions of municipalities and provinces, without working to reduce such a phenomenon despite the promised projects in urban planning, preservation of green spaces and improvement of services. The articles concentrated on the Ghouta region without mentioning Douma in particular, and we have not found any article or reference addressing the effect of urbanization in Ghouta on changes in its cities' traditions, customs and ways of earning a livelihood by inheriting traditional crafts such as the Aghabani industry. The sources also indicated the difficulty of living in Ghouta due to poor housing and water and power services, apart from the possibility that some of its inhabitants might be stripped of their agricultural property for the construction of government facilities.

The third reason for the low rate of employment in the Aghabani industry is the increase in girls' education in the city, where girls were previously prevented from completing school upon reaching a certain age so they could start working in the craft their mothers had taught them since childhood. That situation has changed, and as the percentage of girls who stay in school has increased, that of girls working in the Aghabani industry has dropped.

We could not find any documentation or references regarding the efforts of the Syrian government at that time to support the tangible and intangible cultural heritage in general or the Aghabani industry in particular. The articles we reviewed also did not concentrate on the status of Douma specifically and the development of education there before 2011. For example, an article talked about the phenomenon of children dropping out of school in general without specifying the areas. That article also mentioned the Compulsory Education Law, which requires children to attend school up to the age of 15 years, imposing a financial penalty on parents and holding them subject to prison terms of up to a month if they prevent their children from going to school. The article also stated that female dropout rates are high because of early marriage[89]. Another article, titled "Spending 132 Billion

---

[89] Afra Mohammed, "Syria's Student Dropout Problem: A Phenomenon Exacerbated by Poverty, Customs and Lack of Law Enforcement" [electronic reference], on Deutsche Welle's DW Akademie website, available at https://goo.gl/tuZn1t, published on February 21, 2010, accessed on June 7, 2018.

Syrian Pounds[90] on Education in Syria"[91], reported that enrolment in secondary education reached 77 percent in 2008, without elaborating the percentage of males or females or any details related to the Syrian areas, whether rural or urban. These lacks made the ability of the research to ascertain information related to the high rate of female education in Douma not readily available, especially in the light of the impossibility of access to any documents or official statistics that indicate or confirm this information due to the current situation of the country, the difficulty of communication with these bodies or even having some of those documents lost during the current events.

Some merchants and artisans inherited the work of manufacturing and marketing the Aghabani products from their ancestors, while others learned the industry-associated crafts since childhood according to the customs of sending the children to learn from the sheikh of the craft[92] or based purely on personal desire. Marketing was done through exporting the products to a number of Arab countries such as Kuwait and Morocco, and European countries like France[93].

In the end, the shortage of working women did not result in a significant change in the merchants' business system. Some of the female workers started to deliver the embroidered work to their shops in Damascus instead of the merchants having to go to Douma as before. The merchants would sell part of the goods and store the rest in their warehouses. The work of artisans is not that different from how it was before 2011. However, the protests that broke out in Douma and Ghouta in general, and the subsequent siege, displacement and destruction left their impact on the various aspects of life, including the Aghabani industry. This is what we will discuss in the following paragraphs.

---

[90] Equivalent to US $614,382.
[91] "Spending 132 Billion Syrian Pounds on Education in Syria" [electronic reference], *Zaman Al Wasl* newspaper, available at http://bit.ly/2YncTqS, published on January 20, 2010.
[92] Najat Qassab Hassan, Damascene Discourse: Memoirs 1, 1884–1983, p. 259.
[93] Kayyal Mounir, Damascene Arts and Industries, p. 113.

3. The Impact of War on the Industry

Douma was among the first cities to participate in the peaceful protest movement in 2011, and among the cities that came under the control of the armed opposition[94]. According to the Doumaniya[95], women had participated in the movement's demonstrations and civil activities. When events developed and the peaceful movement devolved into armed conflict, many women lost the men their families had depended on, due to death, arrest or their joining the fighting forces within the armed opposition. These reasons formed an urgent need for women to look for work[96]. The Doumaniya suggested to some of them that they return to work in the Aghabani industry. Having quit the craft for a long time, they did not welcome the idea, but they were persuaded in the end. They arranged their roles so that the Doumaniya's task was to secure materials—like cloth and threads—while the women would do the embroidery. "The women disliked the idea of going back to work in the embroidery of Aghabani, but I insisted that they should prepare the embroidery machines by soaking them in kerosene to remove the rust, and I would bring them the necessary materials," said the Doumaniya[97].

Groups have worked with the women in support projects to help those who have lost their breadwinner to work and secure their daily lives. These projects focused on resources and industries in the same areas as those of the Aghabani industry. In his article titled "Thus the Doumani Women Revived Syria's Heritage", Ziad Ghosn said that the situation in Douma during the war and the difficulty of finding an alternative to female embroidery workers were among the factors that led to the cessation of the industry in the first year after 2011, because of the momentum of incidents in the city, the participation of women

---

[94] "Douma Protests on 'No to Federalism' Friday" [electronic reference], on the *Syria Untold* website, available at https://goo.gl/wRhMXA, published on April 1, 2016, accessed on May 3, 2018.

[95] Lana Mradni (the author), "On the Aghabani Industry in Damascus", an interview with the Doumaniya (a nickname).

[96] Rasha Saleh, "Syrian Women: Suffering and Challenges" [electronic reference], *Al-Nabad*, available at https://goo.gl/kyNxdA, published on April 19, 2017, accessed on May 26, 2018.

[97] Lana Mradni (the author), "On the Aghabani Industry in Damascus", an interview with the Doumaniya (a nickname).

in the peaceful protest movement and the lack of access to the necessary raw materials for embroidery[98]". But this contradicts the information provided by the Doumaniya, who stressed that the Aghabani industry was shrinking 15 years before 2011, for several reasons mentioned in the previous paragraphs[99]. However, the urgent need for work and community initiatives by civil activists interested in the identity-related heritage helped the industry to recover despite the previous difficulties or those resulting from the war. This can be summarised as a series of specific factors such as the difficulty of providing raw materials, a shortage in the labour force and the challenges of marketing and sales. This is what we will review in detail in the following paragraphs.

A) The Provision of Raw Materials

The Aghabani industry depends on the availability of numerous raw materials, including the embroidery threads, the cloth to be embroidered, the embroidery sewing machines and machine oil, as well as materials for printing designs on the cloth in the pre-embroidery stage. Despite the many power and Internet outages in Douma, the women were able to continue the work of the Aghabani embroidery using machines without electricity and an old stock of yarns and cloth. While the merchants continued to get the cloth out of the stock held by the state in its factories, as the finest types of cloth were the ones that were previously manufactured in public sector factories such as al-Khomassya company, and the General Company of Maghzel and Manasej (Weaving and Textiles) in Qaboun and al-Dibs in Sahnaya, most of which were located in active combat areas and thus they stopped working for years. But the stock was still enough to meet the demand.

An article in *Tishreen*, an official newspaper, reported that the General Maghazel Company was "exposed to terrorist attacks on a daily basis, pushing the company to halt." However, the company resumed its production, according to the same article, published on October 27,

---

[98] Ziad Ghosn, "Thus the Doumani Women Revived Syria's Heritage" [electronic reference], *Al-Akhbar* newspaper, available at https://goo.gl/s8EFaM, published on May 22, 2017, accessed on May 23, 2018.
[99] See the second section of Part 1, "The Aghabani Before 2011".

2016[100], which stated that the public company's sales in 2014 hit 480 million Syrian pounds, equivalent to approximately 2.234 million US dollars. In 2015, sales reached about 400 million Syrian pounds, equivalent to approximately 1.861 million US dollars, and in 2016 sales were at about 375 million pounds, or approximately 1.745 million US dollars, at the time of writing the article.

The *Tishreen* article's Information is contradicted by information received from members of the research sample, who confirmed the factories were not operating at the time they were interviewed. One of the respondents emphasised the need for the state to support them to re-operate the factory to secure the cloth, and expressed his resentment of the pressures imposed on them through the work of the Ministry of Interior Affairs and the Consumer Protection Committee, which control the prices of goods in the market and oblige each importer, producer, merchant, worker in the industry or trade or stakeholder of any commodity to provide a declaration on the materials and commodities they possess[101]. The cost of each piece is checked without taking into account the continuous change in the market in terms of the availability and price of raw materials, and the change in the cost of production of an Aghabani piece from one stage to another. These concerns were mentioned in the context of interviews about the existence of differences in the pricing of goods produced in the market and those that are marketed by some projects or official or semi-official institutions, which will be considered in detail in the second part of the research. The goods are sold at expensive prices that do not correspond to the cost of production, yet the Consumer Protection Committee of the Ministry of Interior Affairs do not interfere to reduce it.

As for the embroidery threads or yarns, importation was possible but the source of imports changed because of the sanctions imposed by Europe, the United States and some Arab countries on trade with Syria, which led to a decline in foreign trade and a rise in prices, ac-

---

[100] Hanawi Maada, "With Its Workers' Expertise and Its Own Efforts: The General Maghazel Company Resumes Production" [electronic reference], Tishreen newspaper, available at https://goo.gl/pQ41zb, published on October 27, 2016, accessed on June 7, 2018.

[101] Syrian Arab Republic: Syrian Prime Ministry, Law No. 14 of 2015 on Internal Trade and Consumer Protection.

cording to an article by Batoul Abdo *Al-Thawra*, another of Syria's official newspapers, titled, "Syrian Exports: Crisis Scenarios and Implications"[102]. Table 2 shows the different sources of two raw materials, threads and cloth, before and after 2011, based on information we got from merchants and members of the research sample.

Table 2: Sources of Raw Materials Before and After 2011

The Material	Before 2011	After 2011
Threads	Japanese and Eastern European imports and US stocks were available in the local market	Imports from China and India are available in the local market
Cloth	State factories	What remains of state factories' stock

On the other hand, materials for the printing phase of the Aghabani production were still available, but they became quite expensive. The problem was the availability of the wooden printing moulds. There were no woodworkers to manufacture such moulds. The printing worker in the research sample told us that he kept a complete set of engraving moulds in case he needed them in the future, as he had not found anyone to buy them[103] (See Photo 7).

---

[102] Batoul Abdo, "Syrian Exports: Crisis Scenarios and Implications" [electronic reference], *Al-Thawra* newspaper, available at https://goo.gl/iYpmBE, published on March 20, 2012, accessed on June 6, 2018.

[103] Lana Mradni (the author), "On the Aghabani Industry in Damascus", an interview with Abu Abdo, the printing worker (a nickname), (n.p.), Damascus, on March 23, 2018.

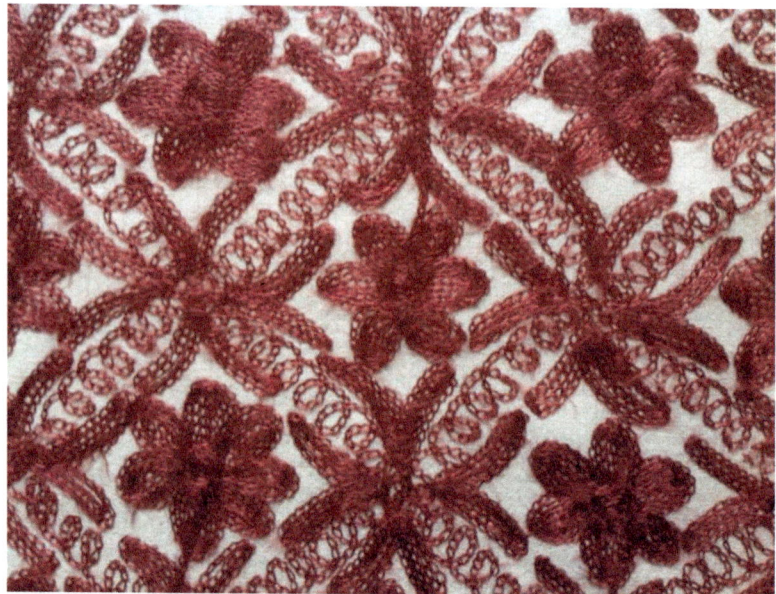

Photo 7: A form of Aghabani's Embroidery - The Dhamma Design – from Khouyout Alamal of the - Aghabani[104]

Similar conditions also apply to the replacement parts of embroidery machines, whose prices rose, if available, as did the prices of the oils needed for machines[105].

B) The Labour Force

The Aghabani is made in stages, starting with cloth processing and printing, the embroidery, cleaning and ironing the finished pieces, and finally selling the final product in the markets. At the beginning of 2011, communication between the Aghabani merchants and most of the female embroidery workers was terminated due to the security situation, as the city of Douma was witnessing almost daily demonstrations against the Syrian government and fighting between the armed opposition and the government was escalating. The government re-

---

[104] From the "Threads of Hope – Aghabani" or "Khouyout Alamal – Aghbany Facebook page at https://goo.gl/7BwM3i.

[105] Lana Mradni (the author), "On the Aghabani Industry in Damascus", an interview with Abu Saeed, the machine repairman (a nickname), (n.p.), Damascus, on April 23, 2018.

sponded to the demonstrations with shelling and tanks, displacing a large number of residents of Douma to Damascus. Then a group of female workers displaced in Damascus contacted the merchants to offer them working in embroidery.

Some of the female workers managed to carry the machines in a cloth package called a "buqcha" on their heads as they came out of Douma under the bombardment, for the machines were "more important than their children"[106], one of the sample members said. Other working women were unable to bring the machines with them at the time of displacement. The alternative was to rely on used French machines in Damascus, which needed spare parts, a circumstance that added great difficulties for some workers, in addition to the security constraints regarding obtaining permission[107] to rent apartments in Damascus and the neighbours' complaints of the loud sound of machines, which sometimes led to the expulsion of the entire family from the house[108].

Merchants continued to work, relying on the Aghabani stocks they kept in their shops' warehouses, in addition to pieces produced by the small number of female workers who were displaced to Damascus. Work and even limited production were facilitated by the availability of printing, washing and ironing services provided by some craftsmen who continued to work despite the low market demand[109].

The experience of the embroidery machine repairman, Abu Saeed, a member of the research sample, for example, reflects the experience of displacement of some male and female workers. Abu Saeed is from Douma, and he stayed there until he left two years after the Syrian government began its siege of the city in late 2013[110]. He stopped

---

[106] Lana Mradni (the author), "On the Aghabani Industry in Damascus", an interview with the merchant Abu Samer (a nickname).
[107] Rasha Saleh, "Syrian Women: Suffering and Challenges".
[108] Lana Mradni (the author), "On the Aghabani Industry in Damascus", an interview with the Doumaniya (a nickname).
[109] The printing worker stopped working for two years and the embroidery worker for four years.
[110] Sulafa Jabbour, "Damascus' Eastern Ghouta: A Hundred Days under Siege" [electronic reference], *Al Jazeera*, located at https://goo.gl/gcphsR, published on January 12, 2014, accessed on June 1, 2018.

working when he was asked to work in weapons manufacturing and maintenance until he was able to move to Damascus. There, he had to stay at home for a while to take care of his ill wife before he returned to work in his job anew as an artisan at the workshop of one of his father's former students. Then, he got a loan from a non-governmental organization and opened his own workshop again.

As for printing mould makers, there were very few of them in Damascus. We were unable to meet the only mould maker for whom the rest of the sample could give us an address, and who used to work in the woodwork of moulds for entertainment, as he owned a television agency. However, the information we got about the moulds was contradictory; some said the moulds remain usable for two or three years, while others said they remain usable for 10 years and then are sold off to foreign tourists as souvenirs. (See Photo 8)

Photo 8: Aghabani Designs' Printing Moulds, "Threads of Hope – Khouyout Alamal – Aghabany"[111]

C) Internal and External Marketing of Products

The Aghabani products are mainly sold internally—in Damascus, it is hard to find a house (see Photo 9) without an Aghabani product—besides exporting them to external markets. However, since before

---

[111] From the "Threads of Hope – Aghabani" or "Khouyout Alamal – Aghbany Facebook page at https://goo.gl/V6JbeM.

2011, the Aghabani industry has faced competition from imported products. After the demand greatly decreased due to war conditions, the competition between these products became greater. The market was limited to the internal dispersal of products in the form of gifts given by the Chamber of Industry to the diplomatic delegations coming to the country to negotiate. Recently, with the Russian presence in Syria, the sale to the Russian community became a source of marketing and selling the products. Syrians residing abroad also assisted in marketing and selling, as well as participating in the few fairs held in the Gulf region.

Photo 9: an Aghabani merchant inside his shop - by Maher Al Mounes - Damascus - 2018

The war has greatly affected people's confidence in the quality of Syrian handicraft products because of some attempts by some countries—such as China, Bulgaria, some Gulf countries and other neighbouring countries—to reproduce traditional products and sell these fake copies in international markets as genuine parts of Syrian heritage[112]. Abu Muhammad, the embroidery worker, explained his experience in opening a workshop in Algeria, where he encountered difficul-

---

[112] Bassam al-Mustafa, "Our Traditional Craft Industry and the Dangers of Its Counterfeit?!" [electronic reference], *Al Azmenah* magazine, available at https://goo.gl/ATjGpg, published on December 20, 2012, accessed on August 29, 2018.

ties in gaining the trust of customers because of their insistence on buying goods with labels on which it was specifically written "Made in Syria" because of their trust in the quality and taste of Syrian handicrafts[113]. "The lack of a trademark and the Syrian government's failure to register traditional products as national products prevent the development of these crafts through the use of labour standards at the global level"[114]. Moreover, the lack of documentation on the basis of modern principles through the development of a promotion strategy facilitates the process of fraud and theft done by individuals or states.

This section presents some of the changes in industry conditions as a result of the war, which provide some features, though not an overall picture, of the situation of the Aghabani industry after 2011, but it highlights different aspects of the responses of stakeholders in these circumstances, both workers and merchants, or actions taken by the Syrian government itself, affiliated bodies or other independent entities. This paves the way for the next section, which further explores these responses in order to find out more about the current status of the Aghabani industry from the principle of documentation and knowledge of the motivations of workers/stakeholders and the extent of impact of these conditions on them.

---

[113] Lana Mradni (the author), "On the Aghabani Industry in Damascus", an interview with Abu Muhammad, the embroidery worker (a nickname), (n.p.), Damascus, on April 23, 2018.

[114] Bassam al-Mustafa, "Our Traditional Craft Industry and the Dangers of its Counterfeit?!".

Part Two –
Response Mechanisms to the Risks and Challenges Facing the Aghabani Industry

This section summarises the various risks threatening the industry as a result of the war in a continuation of the attempt to document the industry's status and development after 2011. It also discusses the responses of industry stakeholders such as governmental and non-governmental organizations, UNESCO and industry workers to find solutions to these conditions and risks. It also aims at monitoring the impact of war on industry stakeholders and workers, and finally proposes a set of recommendations to help the Aghabani industry community to continue its work and production.

It should be noted that the risks presented in this research may be an inevitable result of war-related changes as well as accumulations of societal and non-supporting policies for the industry since before 2011 and after. The manufacturing of wooden printing moulds, for example, is seriously threatened by extinction. Members of the research sample informed us that the only mould manufacturer in Damascus was an ill elderly man, and his shop was closed every time we tried to visit it. There is no information about the existence of another mould manufacturer who can transmit the work for generations to come, and there is no possibility of replacing the moulds with modern technology. The solution was to inherit the moulds within the family and store the spare moulds at houses. The printing worker in our research sample even expressed his desire to learn the mould engraving industry if necessary.

Despite the fact that Damascus University's Faculty of Fine Arts annually graduates dozens of wood engravers, there is no encouragement or interest in supplying traditional industries with educated cadres. Young people, the embroidery worker said, are reluctant to learn such industries. He considered this to be a real problem. "For the past ten years, there is no single new young man in this craftsmanship"[115], he said. Others also indicated that the transmission of the industry to future generations is threatened, even if only partially, as half of the research sample of merchants and craftsmen said they were transmitting the work to their sons, while the other half said they could not due to their sons' lack of interest, death or moving abroad. The merchant Abu Bassam works in this industry with most of his family members, i.e. his brothers and cousins, while the sons of the merchant Abu Samer did not care to work in the industry and left their father to work alone[116]. One of the craftsmen, the printing worker, said that he did not want his son to work in the industry, for others do not appreciate it and prefer modern machine embroidery over handicrafts because of its speed and cheap prices[117]. As for the machine repairman, his son will work with him in using the lathe after graduating from Damascus University's Faculty of Mechanical Engineering to develop the work by integrating science with inherited experience[118].

The high proportion of Syrian youths who are at military recruitment age and who want to travel to neighbouring countries out of fear of being forced to join the military service is one of the fundamental reasons leading to a decrease in the number of males in the labour market who are interested in learning skills needed in the handicrafts industry. This has been reported in many news articles, as in an *Asharq Al-Awsat* article that notes the beginning of a new wave of youths seeking to flee from Syria that coincided with the Syrian government's launching of a new campaign to force the remaining civilian youths into the mandatory and reserve military services[119].

---

[115] Lana Mradni (the author), "On the Aghabani Industry in Damascus", an interview with Abu Muhammad, the embroidery worker (a nickname).
[116] Lana Mradni (the author), "On the Aghabani Industry in Damascus", an interview with the merchant Abu Samer (a nickname).
[117] Lana Mradni (the author), "On the Aghabani Industry in Damascus", an interview with Abu Abdo, the printing worker (a nickname).
[118] Lana Mradni (the author), "On the Aghabani Industry in Damascus", an interview with Abu Saeed, the machine repairman (a nickname).
[119] "A Recruitment Campaign for the Remaining Young Syrians Forcing Them to Emigrate" [electronic reference], *Asharq Al-Awsat* newspaper, available at https://goo.gl/fcQA4A, published on December 21, 2016, accessed on May 3, 2018.

In addition to the above, the competition from machines using modern technologies represents a threat to traditional handicrafts amid the lack of protection policies, especially because they reduce time and production costs. Previously, a tourist could distinguish the handmade product from the machine-made ones by the small defects that lend an aesthetic character to handmade items, as the machines do not make mistakes. Accordingly, the buyer would choose the handmade product with its imperfections regardless of its price. Now, the consumer is looking for the product's function rather than its aesthetic aspect, so he or she is ready to buy the machine-made product as long as it meets the purpose at cheaper prices. In other words, "the customer wants what can cover him rather than something to admire and watch!"[120], according to Abu Muhammad, the embroidery worker. The low purchasing power of Syrian customers also leads to their reluctance to purchase traditional handicrafts, which they consider to be luxuries.

These difficulties did not naturally lead to the disappearance of the industry, however, as some official and non-official projects and initiatives continued an attempt to operate and preserve the industry from extinction, according to the various groups' policies and interests. Many entities have played a role in this. These can be summarised as official and non-official institutions, professional associations, UNESCO and industry workers. We will review the role of each of them in detail in the following paragraphs.

1. The Role of Governmental and Non-Governmental Institutions

In 2005, the Syrian government ratified the 2003 United Nations Educational, Scientific and Cultural Organization (UNESCO) Convention for the Safeguarding of the Intangible Cultural Heritage. The first section of this research (the Aghabani before 2011) discussed some of the reasons for the decline in the Aghabani work over the 20 years before 2011 and the lack of the government's role in the protection of intangible heritage, which has led to a decline in crafts and has driven

---

[120] Lana Mradni (the author), "On the Aghabani Industry in Damascus", an interview with Abu Muhammad, the embroidery worker (a nickname).

craftsmen to other professions, as well as other consequences, like the opening of markets for imported industrial products and their competition in the market with their low prices; the imposition of taxes not commensurate with the economic income of craftsmen who faced other challenges related to higher production costs and marketing difficulties; the lack of laws protecting the local industry and traditional crafts, which led other countries to reproduce and sell copies at competitive prices in both domestic and global markets; and administrative corruption, which fostered the spread of random urbanization at the expense of the old urban style of Arab houses, which often consist of two floors inhabited by a single family[121]. The latter may have played a role in reducing the work in the Aghabani industry because of the proximity of homes in the new vertical buildings and the annoyance of neighbours from the machines' noise and their complaints against women working in embroidery. Finally, the research could not manage to reach any governmental or academic information that illustrates how the government's response to social changes affected, in one way or another, Douma's intangible cultural heritage as the lack of female workers increased due to the high rate of education in Douma in particular or in Syria's major cities and rural areas.

In the meantime, the Syria Trust for Development was established in 2007 and is chaired by Asma al-Assad, Syria's first lady and the wife of Syrian President Bashar al-Assad. The Trust defines itself as a "leading field development organization in social change in Syria, based on agreements and partnerships with governmental institutions, national and international organizations"[122]. It is also considered one of the most important players in the field of preservation of the Syrian intangible cultural heritage. In 2013, it launched the Syrian Handicrafts company within its programmes. "It is the first social institution working to document and promote the intangible cultural heritage while improving the conditions of craftsmen and local communities and seeking to institutionalise their practices in order to ensure the sus-

---

[121] Yasmin Jani, "The Status of Ancient Arab Houses to the Syrians", *Zaiton* magazine, available at http://www.bit.ly/2UjfPHa, published on July 22, 2017, accessed on August 12, 2018.

[122] From the official website of the Syria Trust for Development, available at https://bit.ly/2OOMsq4, accessed on April 5, 2019.

tainability of the Syrian heritage and its continuity effectively," the handicrafts company said on its website[123]. Its first brand, Ubbaha, is the first registered trademark of handicrafts products.

During a workshop on the reality of traditional industries held in Damascus in 2016, Syrian Minister of Tourism Bishr Yazigi praised the role of the Syria Trust for Development in addressing the threats to the handicrafts industry[124]. In 2015, the Trust, together with the Syrian Handicrafts Company, opened the Ubbaha project to manufacture, sell and modernise traditional handicrafts to support the affected craftsmen and create new jobs for them[125]. The project "enabled the activation of 14 crafts benefiting 142 craftsmen who had made 222,000 products, which in 2015 amounted to more than 42.5 million Syrian pounds," equivalent to about 225,000 US dollars in 2015[126]. However, the article provides no information on the mechanism of action, sale or how to distribute income to the craftsmen. Neither the Ubbaha project's official website nor social media provides any information on the prices of products for which Ubbaha has chosen a number of shops in al-Takiyya al-Sulaymaniyah (see Photo 10) to display their products through an agreement with the Ministry of Tourism[127]. An article on the *New Arab* website by Abdul Razzak Diab, titled "al-Takkiyya al-Sulaymaniyah in Damascus", stated that Ubbaha rehabilitated the shops with funding from their owners. It also imposed fines and renovation fees and added large amounts to the rents for the shops, which

---

[123] From the official website of the Syrian Crafts Company, available at https://bit.ly/2UiFsrB, accessed on April 5, 2019.

[124] SANA, "A Workshop on the Reality of Traditional Industries" [electronic reference], SANA (Syrian Arab News Agency), available at https://goo.gl/yuBc31, published on August 17, 2016, accessed on May 3, 2018.

[125] "Ubbaha Revives Syrian Handicrafts in al-Takiyya al-Sulaymaniyah" [electronic reference], *Al-Boselh* website, available at https://goo.gl/TfKcfB, published on October 22, 2015, accessed on June 7, 2018.

[126] "Activation of 14 Crafts Benefit 142 Craftsmen with Revenues of 84 Million" [electronic reference], *Al-Baath* newspaper, available at https://goo.gl/1vQXem, published on August 29, 2016, accessed on May 10, 2018.

[127] "Ubbaha Revives Syrian Handicrafts in al-Takiyya al-Sulaymaniyah".

the craftsmen could not pay because of the decline in their sales, forcing many of them to leave the market and travel abroad[128].

Photo 10: Ubbaha Shops in al-Takiyya al-Sulaymaniyah - Facebook Page - Damascus - 2018[129]

In 2014, the Syria Trust for Development issued the first part of a book titled *The Non-Material Heritage: Skills Associated with Traditional Craftsmanship*[130], which documents 100 elements of Syria's intangible cultural heritage in collaboration with the Ministry of Culture and UNESCO's Country Commission for Syria. The book documents the Aghabani[131] as one of these 100 elements in terms of origin, stages of manufacturing and threats to the craft without including any information about the deterioration of the security situation in the cities of

---

[128] Abdul Razzak Diab, "Al-Takiyya Al-Sulaymaniyah in Damascus" [electronic reference], on *The New Arab (Al-Araby Al-Jadeed)* website, available at https://goo.gl/Zb737V, published on February 27, 2018, accessed on September 7, 2018.
[129] From Ubbaha's official Facebook page at http://bit.ly/2FLynX3.
[130] Talal Maallah, The Non-Material Heritage: Skills Associated with Traditional Craftsmanship, pp. 61–63.
[131] Talal Maallah, The Non-Material Heritage: Skills Associated with Traditional Craftsmanship, pp. 61–62.

Damascus and Douma. It states that "the link between the form of the industry's work and workshops makes the ability to continue and withstand the potential challenges resulting from violent changes in the market, the lack of legal frameworks to define and protect the profession, and the lack of organic relations with technical, administrative and economic research"[132]. These constitute real threats to industry, as the process of manufacturing goes through stages of intermittent work at workshops or among individuals dispersed between Damascus and its countryside, specifically the city of Douma, making the fate of these workshops and individuals linked to the place where they are located and where the changes took place, especially in Douma, most of whose people were forced to flee. Consequently, it is impossible to predict the fate of a large number of workers in the industry who remain there or even those displaced to Damascus or Idlib. The words "violent changes" in the text quoted from the book are the only reason given for the possibility of a situation-changing factor generating challenges that the industry's manufacturing model may not be able to deal with. There is no mention of the conflict since 2011 that has killed thousands, forcibly displaced Syrians and uprooted their communities, according to a Sky News article titled, "Douma: A New Chapter of Forced Displacement in Syria"[133]. Thus, the book excluded the risks resulting from war and the displacement of the Aghabani industry's community.

While the Syrian government places the issue of protecting the intangible cultural heritage as one of the priorities for which a range of measures are proposed to mitigate the negative effects on the traditional industries, the conditions for the return of the residents of Douma remain an unresolved issue that may have political dimensions, and thus it has either to submit to government policies or to remain outside its authority. It is not possible to predict the effectiveness of the initiatives taken by governmental or non-governmental institutions aimed at training women from other cities on the embroidery of

---

[132] Talal Maallah, The Non-Material Heritage: Skills Associated with Traditional Craftsmanship, p. 63.

[133] "Douma, A New Chapter of Forced Displacement in Syria" [online reference], *Sky News*, available at https://bit.ly/2OSyYcN, published on April 9, 2018, accessed on September 15, 2018.

the Aghabani and other training courses on project management to get the woman a machine before they stop their intervention[134], according to a question raised by Ziad Ghosn's article[135], as female workers faced social barriers preventing their work outside their homes. Also, it is not possible to know the extent of professionalism of the trainees and their keenness to maintain the aesthetic of the product. No information has been obtained indicating a clear follow-up plan by these institutions. As for the Doumani female workers, it is not possible at this time to predict their fate and whether they would choose to return to Douma after the siege ended and the city returned to the Syrian government's control.

The same applied to the mechanisms of support, sales and price fixing taken by these institutions without dealing directly with the industry's incubating society and the development of solutions by its participation, which is contrary to what Graciela Singer[136] argues, confirming that the intangible cultural heritage cannot be considered without taking into account the individuals and communities that have created it and maintained it through its transfer. It is very likely that the status of the Aghabani craft, as an intangible cultural heritage, will be affected if the Syrian authorities continue to deal with it in the same way, i.e. if the government continues to take measures without consulting the society that manufactures it, including the workers and artisans of Douma, and the craftsmen and merchants residing in Damascus. This is also confirmed by a member of the research sample. By ensuring suitable living conditions for the craftsmen who were forced to leave their city, the rest of the industry actors will be able to function, said the merchant Abu Samer, as "the countryside feeds the city, and is considered the lung by which the country breathes"[137].

A merchant stated that one of the problems plaguing tradesmen and craftsmen today relates to the work of the Ministry of Internal Trade and the Consumer Protection Committee. Supply officials unrea-

---

[134] Lana Mradni (the author), "On the Aghabani Industry in Damascus", an interview with Abu Saeed, the machine repairman, (a nickname).
[135] Ghosn Ziad, "Thus the Doumani Women Revived Syria's Heritage".
[136] Graciela Singer, "The Importance of the Intangible Cultural Heritage".
[137] Lana Mradni (the author), "On the Aghabani Industry in Damascus", an interview with the merchant Abu Samer (a nickname).

sonably check the cost of handmade products that are difficult to estimate in light of the war, where unexpected additional costs might be incurred. For example, it is difficult to follow up the female workers, where some consume more cloth than others, and the search for a particular cloth requires additional effort because of its scarcity in the market, while it used to previously come from the government textile companies[138]. "It is exhausting to respond to the demands of accuracy in the calculation of the cost while you see your own goods sold elsewhere at expensive prices without noticing the Supply Officials objecting to that," said one of the respondents[139]. This might be attributed to the lack of confidence in the transparency of the application of the procedures of consumer protection committees and the predominance of favouritism.

On the other hand, the Syrian Minister of Tourism Bishr Yazigi responded to the artisans' allegations and grievances, blaming them for lack of contribution to the state treasury, even if with a single $1, and deeming the taxes and duties owed to the Ministry of Finance "almost negligible, or even just a shameful figure to mention"[140]. He also complained of "craftsmen's non-compliance in the antique market in al-Takiyya al-Sulaymaniyah, for example, where the monthly rental fees for a shop do not exceed 10,000 Syrian pounds, which is equivalent to 22 US dollars[141], yet artisans refused to repair the façade board of the market"[142]. This raises a question about the ability of artisans, during the time of the search and the war, to pay taxes and the stores' rental fees amid the low purchasing power and demand for traditional products in the market.

It is worth mentioning that a number of local community initiatives have emerged since 2012, such as the "Threads of Hope – Khouyout Alamal – Aghabany" project to protect the industry. Other associations, such as al-Mabarrah, have been providing the displaced

---

[138] Abdul Razzak Diab, "al-Takiyya al-Sulaymaniyah in Damascus".
[139] Lana Mradni (the author), "On the Aghabani Industry in Damascus", an interview with the merchant Abu Samer (a nickname).
[140] "Activation of 14 Crafts Benefit 142 Craftsmen with Revenues of 84 Million".
[141] Exchange rate: US $1 = 445 Syrian pounds on August 5, 2018.
[142] "Activation of 14 Crafts Benefit 142 Craftsmen with Revenues of 84 Million".

Aghabani female workers with machines and helping them to contact commercial projects as a way to help them earn a living.

Since the preservation of the intangible cultural heritage requires a strategy to enhance it and to ensure its constant adaptation to the cultural development of its creators and stakeholders, the Syrian government must create the appropriate conditions for the development of the intangible heritage so as to reduce the internal influences of the dominant sectors and prevent them from affecting the spontaneity of this process[143]. It is likely that the current strategy adopted by the Syrian government, its affiliated institutions such as the Syria Trust and the organizations that control the development of the Aghabani industry through interventions and solutions that do not include the place where the industry originated since the Ottoman era on the one hand, or the participation of the city's community that had been scattered across Syria and in the Diaspora on the other hand, had led, according to Lenzerini[144], to the loss of the originality of the intangible heritage because of its prioritization according to interests beyond those of its creators and stakeholders. Thus, preserving the intangible heritage requires its adaptation to the cultural identity of the communities and individuals concerned by re-creating themselves as a reflection of the cultural and social development of those groups and individuals.

To learn more about the role of the Syrian governmental institutions and organizations in this regard, we will present the role of the General Union of Craftsmen in supporting the industrial community by supporting the industry's workers and their representation during the war and before.

2. The Role of Trade Unions through the General Union of Craftsmen

The General Union of Craftsmen was established in 1969 with the aim of "engaging craftsmen in building the Arab socialist society by organising production and services, increasing and improving their quality, caring for the material, moral, health, cultural and social interests of

---

[143]   Federico Lenzerini, "Intangible Cultural Heritage: The Living Culture of Peoples", pp. 101–103.
[144]   Federico Lenzerini, "Intangible Cultural Heritage: The Living Culture of Peoples".

the craftsmen, raising their living standard, [...] liberating them from being exploited, facilitating the transfer and marketing process, working to raise craftsmanship efficiency, increasing labour productivity, developing skilled craftsmanship, and developing and promoting talented craftsmanship"[145].

In spite of the establishment of the union based on the abovementioned principles, the handicrafts in general suffered from many obstacles that pushed them to decline and toward extinction. The set of decrees and laws that describe the work of artisans and the functions of the union did not take into force that plans and procedures to preserve the crafts being linked to heritage and memory and that form part of Identity can, through development and support, contribute to the development of the economy. Instead, they are all corrupt and bureaucratic policies[146], according to an article by Bashar al-Hajali and others titled, "Traditional Crafts and Industries: A Homeland's Memory. How Can We restore Its Spark?"[147]. Noteworthy is the scarcity of articles published in official newspapers, such as *Al-Thawra Online*, that criticise the efforts of governmental institutions, while there were some articles published in 2011 and 2012, which coincided with the beginning of the movement in Syria. The latter helped us to obtain some information that reflects the reality of crafts in general and the ability to partially analyse the situation of artisans up to that period.

On the other hand, Mohammed Fayyad al-Fayyad praised the work of the General Union of Craftsmen for providing the information and logistical facilities that helped him in compiling the "Encyclopaedia of Traditional Crafts in Syria", issued by the union itself. He considered this effort as one of the most important means of documenting and preserving the crafts heritage from extinction, supported by photos for the various crafts, according to Fadi Telfah's article on eSyria titled "The Encyclopaedia of Traditional Crafts: A Documentation Aimed at Protec-

---

[145] Ammar al-Nahar, "A Study on the Book of Traditional Crafts in Syria", *Journal of Creative Circles*, Third Issue, Damascus University, 2015.

[146] Bashar al-Hajali et al., "Traditional Crafts and Industries: A Homeland's Memory. How Can We Restore Its Spark?" [electronic reference], *Al-Thawra Online*, available at https://goo.gl/7nPGqx, published on December 22, 2012, accessed on September 7, 2018.

[147] Bashar al-Hajali et al., "Traditional Crafts and Industries: A Homeland's Memory".

tion"[148]. Upon reviewing the encyclopaedia, we find it praising the craftsmanship organization established in Syria in 1970 within the process of socialist transformation adopted by the so-called Revolution of the 8th of March, 1963[149]. The encyclopaedia lists the history of the development of the artisanal crafts organization and the functions of the General Union of Peasants without addressing the post-2011 period, using old language and terminology (for example, the rehabilitation of women and the disabled)[150]. The percentages and figures are limited to old historical periods[151] and have not kept pace with current changes, which brings into question the applicability of its information. In addition, the encyclopaedia was not mentioned by the sample members we interviewed, which raises the question of whether it is a reference to the work of organizations currently active on the ground.

In 2016 another article in *Al-Thawra Online*, titled "The Union of Craftsmen Adopted a Brand for Globally Desired Crafts"[152], confirmed that the union is interested in all trades, but that the circumstances require that it prioritise some of them, such as the brocade and the Aghabani industries. The president of the General Union of Craft Associations, Yassin Sayed Hassan, believes the best step to maintain the industry can be achieved by training new craftsmen by the sheikh of the craft[153] and transferring these skills to new generations. These actions come in response to the crisis affecting the artisanal sector as well as other sectors. Mr. Hassan said the union has helped by opening

---

[148] Fadi Telfah, "The Encyclopedia of Traditional Crafts: A Documentation Aimed at Protection" [electronic reference], eSyria website, available at https://goo.gl/BtQdjj, published on June 11, 2012, accessed on September 7, 2018.

[149] Mohammed Fayyad al-Fayyad and Majed Hashim Hammoud, *Traditional Crafts in Syria*, pp. 391–395.

[150] Mohammed Fayyad al-Fayyad and Majed Hashim Hammoud, *Traditional Crafts in Syria*, pp. 412–413.

[151] Mohammed Fayyad al-Fayyad and Majed Hashim Hammoud, *Traditional Crafts in Syria*, p. 411.

[152] "The Union of Craftsmen Adopted a Brand for Globally Desired Crafts" [electronic reference], *Al-Thawra Online*, available at https://bit.ly/2KLQtgB, published on December 12, 2016, accessed on September 7, 2018.

[153] The Sheikh of the Craft: An old term that was used in Damascus and Aleppo to refer to the chief of the profession or career. He is the most informed and oldest among artisans and the one who represents them in front of other professions, the state and the Sheikh of the Sheikhs of their craft.

up new markets in friendly countries such as China and Russia and by providing lands for artisanal zones where artisans displaced from active combat zones can work.

The research has not been able to know and understand the crafts' organization and its institutions, and the Union of Craft Associations' different functions and roles – nor the effectiveness of the work of the General Union of Craftsmen or the feasibility of the steps taken in response to the effects of war and the maintenance of handicrafts. This is attributed to the difficulty of monitoring this information and not including a set of appropriate questions in the questionnaires given to the sample members, marketing and expanding their expertise, given that the female workers are not registered in the union, which is exclusive to the artisans working in printing, machine repairs and trade, since the membership is not mandatory but an optional issue, according to the article "Traditional Crafts and Industries: A Homeland's Memory. How Can We Restore Its Spark?"[154]. Thus it remains an open question: Are the artisanal societies involved in making decisions, whether they are related to the creation of suitable environments for their development or to the more efficient means of transferring their crafts to future generations, based on their own interests first rather than the interests of the authority? In order to try to answer this and other questions that seek to explore the roles and responsibilities that must be established to protect the intangible cultural heritage in general and the Aghabani industry in particular, the following section presents some points that focus on the role of UNESCO and its response to the threats that beset the Syrian intangible cultural heritage, especially as UNESCO considers its safeguarding as one of its priorities[155].

---

[154] Bashar al-Hajali et al., "Traditional Crafts and Industries: A Homeland's Memory. How Can We Restore Its Spark?"

[155] UNESCO, "The Intangible Cultural Heritage" [electronic reference], on the website of UNESCO's Cairo Office, available at: https://goo.gl/abeafB, [n.p.], accessed on April 3, 2019.

3. The Role of the UNESCO

In this section, we review some of the main points of the UNESCO Convention for the Safeguarding of the Intangible Cultural Heritage[156], which was ratified by the Syrian government, thus making the terms of the convention binding on the government as a globally ratified agreement that emphasises the importance of safeguarding the intangible cultural heritage as a basic human right and that considers harm affecting any people's cultural and intangible heritage a harm to the cultural heritage of all humanity. In this light, we will try to highlight the role of UNESCO in the conflict in Syria and its statements based on monitoring the impact of the military conflict. Then, we will address some of the initiatives undertaken by UNESCO in response to the danger, such as the programme of "Safeguarding the Heritage", which resulted in the establishment of the Observatory for the Safeguarding of Cultural Heritage[157] in order to mitigate these effects and oblige the conflicting parties to stick to the international treaties. Finally, this role will be analysed by discussing UNESCO's relationship with the Syrian government and the extent of applying the role of supervisor on the ways by which the government responds to the challenges of war on the intangible cultural heritage.

UNESCO has defined intangible cultural heritage as "the practices, representations, expressions, knowledge, skills—as well as the instruments, objects, artefacts and cultural spaces associated therewith—that communities, groups and, in some cases, individuals recognise as part of their cultural heritage. This intangible cultural heritage, transmitted from generation to generation, is constantly recreated by communities and groups in response to their environment, their interaction with nature and their history, and provides them with a sense of identity and continuity, thus promoting respect for cultural diversity

---

[156] UNESCO, "Text of the Convention for the Safeguarding of the Intangible Cultural Heritage, Article 2: Definitions", adopted in Paris on October 17, 2003, available on the UNESCO Intangible Cultural Heritage website at https://bit.ly/2OKUobK, accessed on April 3, 2019.

[157] UNESCO, "Text of the Convention for the Safeguarding of the Intangible Cultural Heritage, Article 2".

and human creativity"[158]. Therefore, the protection and development of intangible cultural heritage enhances the connection and sense of cultural identity defined by Susan Healey[159] as "a sense of belonging that derives from shared origins or characteristics." Shared origins, she explains, "include having the same parents or being born in the same geographical area. Shared characteristics include a common language or livelihood"[160]. UNESCO stressed that "intangible cultural heritage is compatible with existing international human rights instruments, as well as with the requirements of mutual respect among communities, groups and individuals, and of sustainable development"[161].

The Convention for the Safeguarding of the Intangible Cultural Heritage defines safeguarding as "measures aimed at ensuring the viability of the intangible cultural heritage, including the identification, documentation, research, preservation, protection, promotion, enhancement, transmission, particularly through formal and non-formal education, as well as the revitalization of the various aspects of such heritage".[162] Accordingly, the convention obliges the Syrian government to take all possible measures to safeguard the intangible cultural heritage and to ensure its unimpeded transmission to other generations.

UNESCO has made repeated appeals and its former director-general, Irina Bokova, appealed to all conflicting parties in Syria "to safeguard the country's cultural heritage and to take all possible measures to prevent further destruction"[163]. A UNESCO report titled

---

[158] UNESCO, "Text of the Convention for the Safeguarding of the Intangible Cultural Heritage, Article 2".

[159] Susan Healey is a multidisciplinary graduate with a degree in social sciences and has a variety of international experience in policy development, planning, project management, programme evaluation, and organizational change.

[160] Susan Healey, "Cultural Resilience, Identity and the Restructuring of Political Power in Bolivia", a paper submitted for the 11th biennial conference of the International Association for the Study of Common Property, Bali, Indonesia, June 19–23, 2006, p. 4.

[161] UNESCO, "Text of the Convention for the Safeguarding of the Intangible Cultural Heritage, Article 2".

[162] UNESCO, "Text of the Convention for the Safeguarding of the Intangible Cultural Heritage, Article 2".

[163] UNESCO, "'Stop the Destruction of Syrian Cultural heritage!' Urges UNESCO Director-General" [electronic reference], available at http://bit.ly/2JhDNdg, published on August 29, 2013, accessed on April 4, 2019.

"The Intangible Heritage" said that conflict has affected all aspects of cultural heritage, as it had "significant impact on the bearers and practitioners of intangible cultural heritage expressions and on the viability and transmission of practices and know-how of living heritage, essential to maintaining cultural diversity, social cohesion and dialogue between communities"[164]. The same report states that the impact of the conflict on crafts production widened as far as Damascus, although the capital was not directly affected by the conflict, compared to surrounding cities such as Douma or far away ones like Aleppo.

Funded by the European Union and in cooperation with UNESCO, the Urgent Safeguarding of the Syrian Intangible Heritage Project was established in 2014. It launched initiatives aimed at "contributing to social reintegration, stability and sustainable development by protecting the Syrian cultural heritage and safeguarding it from ongoing destruction and increasing losses affecting Syria's rich heritage"[165]. Accordingly, it was decided at an international meeting of experts in 2014 to establish an observatory "to monitor the built, movable and intangible heritage in Syria." The meeting, held at UNESCO's headquarters in Paris on May 26 through 28, 2014, brought together more than 120 experts from 22 countries, including "cultural heritage specialists from Syria and the Syrian diaspora, representatives of Syrian non-governmental organizations," and others, UNESCO said in a news release[166]. It acknowledged that "the intangible heritage in Syria is severely damaged because of social dispersion, and factors related to displacement and emigration. In the ancient city of Aleppo, which has been subjected to the most severe forms of destruction, artisans have

---

[164] UNESCO, "The Intangible Heritage" [electronic reference], UNESCO Syrian Cultural Heritage Observatory, available at https://goo.gl/3b8Uog, (n.p.), accessed on April 4, 2019.

[165] UNESCO, "The Urgent Safeguarding of the Syrian Intangible Cultural Heritage" [electronic reference], UNESCO Syrian Cultural Heritage Observatory, available at https://goo.gl/HjdcUH, (n.p.), accessed on April 6, 2019.

[166] United Nations, "Syria: UNESCO Establishes an Observatory for the Safeguarding of Cultural Heritage" [electronic reference], UN news website, available at https://goo.gl/yBPT5e, published on May 29, 2014, accessed on April 3, 2019.

lost a large number of their workshops, tools and materials. Activities related to the transmission of their skills have also been suspended"[167].

On the other hand, UNESCO has stated that it appreciates the role of the Syrian government in recent years regarding the protection of the intangible cultural heritage, when Damascus applied to join UNESCO's Creative Cities Network and the Syria Trust for Development, headed by Asma al-Assad, was accredited as an advisor in the Committee on the Evaluation of Elements of the World Intangible Heritage (2014–2016)[168]. This highlights some of the contradictions in the role of UNESCO and its relationship with the Syrian government and the Syria Trust for Development, given that the government is a party in the conflict and bears a large part of the responsibility for safeguarding the cultural heritage, as stated in an article titled "Neither the People nor the Monuments: Palmyra and the War Tragedies of the Ruins" by Basileus Zeno[169]. The government also bears responsibility for the displacement of thousands of residents of Douma to Damascus or Idlib, as stated in an article on the *Arab48* news website titled "A Mass Exodus from Douma and Thousands of People Leaving Ghouta for Idlib"[170].

At the same time, the Syria Trust for Development has shifted away from neutrality when it moved to transform some of its programmes after 2011 into programmes of a military character, according to an *Al-Jumhuriya*[171] article that mentioned a "cooperation with the Syrian government to allocate 10 billion Syrian pounds to the project of the homeland's wounded"[172], which includes "the paramilitary martyrs and

---

[167] United Nations, "Syria: UNESCO Establishes an Observatory for the Safeguarding of Cultural Heritage".

[168] UNESCO, "Report by a Non-Governmental Organization Accredited to Act in an Advisory Capacity to the Committee on Its Contribution to the Implementation of the Convention" [electronic reference], UNESCO Intangible Cultural Heritage website, available at https://goo.gl/EBkaas, published on January 18, 2017.

[169] Basileus Zeno, "Neither the People nor the Monuments: Palmyra and the War Tragedies of on the Ruins" [electronic reference], *Al-Akhbar* newspaper, available at https://goo.gl/ffcyJn, published on June 1, 2015, accessed on June 3, 2018.

[170] "A Mass Exodus from Douma and Thousands of People Leaving Ghouta for Idlib".

[171] Karam Mansour, "The Syria Trust for Development: From the Civil Face to the Military Uniform" [electronic reference], on the *Al-Jumhuriya* website, available at https://goo.gl/ESWfwM, published on August 14, 2017, accessed on June 25, 2018.

[172] "The Government, in Cooperation with the Syria Trust for Development, Has Allocated 10 Billion Syrian Pounds to the Project of the Homeland's Wounded" [elec-

wounded and all those who had fought beside the army," according to a Syria Steps article[173], and to the "Laurel Wreath programme to support 100 high school graduates whose relatives were martyrs or wounded within the army and armed forces, including the Internal Security Forces and the police"[174]. This illustrates how the Trust's activities relate to the policies of the Syrian government in the design of programmes and even speech[175]. Thus, the Syria Trust is biased toward one of the conflict parties causing destruction to the tangible and intangible cultural heritage.

This raises a fundamental question regarding UNESCO's mechanisms to monitor governments and institutions that are active in safeguarding the intangible cultural heritage and the extent to which they achieve neutrality, transparency and equality, and whether they design their programmes and identify their responses according to the traditional industries' priorities and benefits, or in accordance with the political agendas imposed by the major dominant countries. Accordingly, what are the criteria UNESCO adopts to assess the role of governments or to designate an institution as an international arbitrator within the Committee on the Evaluation of Elements of the World Intangible Cultural Heritage? And how do these standards correspond to the interest of the industrial society and the support of its workers, especially in conditions of war and the difficulties and risks it entails?

In order to understand the situation of the workers in the Aghabani industry, the following section of the paper provides an account of their reactions and responses to the impacts on the Aghabani industry

---

tronic reference], on the Syria Steps website, available at https://bit.ly/2WWuhBG, accessed on June 25, 2018.

[173] "The Government, in Cooperation with the Syria Trust for Development, Has Allocated 10 Billion Syrian Pounds to the Project of the Homeland's Wounded".

[174] "The Syria Trust for Development Has Launched the Project of the Laurel Wreath" [electronic reference], Tishreen University website, available at https://goo.gl/K65chx, (n.p.), accessed on May 3, 2018.

[175] In a video on YouTube, Asma al-Assad, head of the Syria Trust for Development and wife of Syrian President Bashar al-Assad, is seen speaking in a meeting with employees of the Trust, in which she encourages the values of dialogue and action for the benefit of all the Syrian people without discrimination. The video is available at https://goo.gl/83cb7Y, published on July 22, 2017, accessed on May 3, 2018.
In another video, the first lady praises the women who fought with the government army. That video, titled "Fire Braids", is available at https://goo.gl/5D86y, published on March 22, 2018, accessed on May 3, 2018.

over the past seven years of war. This is a continuation of the research analysis to these motives with the aim of clarifying whether we may classify them under the title of cultural resistance.

4. The Role of Workers in the Industry

Facing all the above, the research raises a question about the change in the awareness and reactions of the industry's workers toward the challenges facing the industry, and whether the workers have the motivation to find solutions to meet the changes the industry is undergoing according to the principle of cultural resilience. This section will present and discuss a set of information provided by the research sample, which is a model for the society of the Aghabani industry in Damascus, to complete the image of the industry's current situation, and to propose a set of recommendations that must be worked on by governmental institutions and affiliated associations, independent cultural institutions and every actor in the industry.

It is worth mentioning that the sample's members unanimously expressed their desire to continue the industry being linked to them as a creative folk craft, in defiance of the various drawbacks left by the war, including death and siege conditions. This section of the research will provide information that illustrates the circumstances experienced by such workers and quotations showing their attitudes toward the industry.

The history of the industry's development confirms the ability of individuals and the Aghabani community to develop and expand its use to include a greater variety of products, ranging from the Ottoman headdress to modern-day fashion and handbags. This is an indication of creativity and determination, according to one of the respondents, who stressed the ability of people working in industry to do miracles. "At first they made the turbans, then the belts and then the tablecloth and jalabiya (garments)," he said. "They continue to innovate and create, for we are making miracles"[176] (see Photo 11). Therefore, the current conditions imposed by the war did not discourage them from keeping going. Despite the challenges represented by the displacement of workers from Douma and the lack of merchants' ability to

---

[176] Lana Mradni (the author), "On the Aghabani Industry in Damascus", an interview with Abu Saeed, the machine repairman (a nickname).

communicate with them, in addition to the challenges of marketing the products internally and externally, the motivation of industry workers to continue will enable them to maintain their craft, as the merchant Abu Bassam asserts. "The negative factors are related to the conditions of labour migration and the lack of marketing," he said. "The positive aspects have to do with our insistence on preserving this heritage"[177].

Photo 11: A picture of a product, from the "Threads of Hope – Khouyout Alamal – Aghabany" Facebook page[178]

---

[177] Lana Mradni (the author), "On the Aghabani Industry in Damascus", an interview with the merchant Abu Bassam (a nickname).

[178] From the "Threads of Hope—Khouyout Alamal—Aghabany" Facebook page: https://goo.gl/7BwM3i

The Impact of War on the Aghabani Textile Industry

The war circumstances were the main motive behind many workers' return to the Aghabani industry after quitting it, as mentioned above. The machines were neglected for a long time and became rusty. When some women in Douma decided to go back to work in the Aghabani industry, they soaked the machines with kerosene oil to remove the rust and began experiments by embroidering tablecloths. Their neighbours admired the idea and joined them to expand the initiative. These motivated other craftsmen, whose trades and crafts relate to the industry, to resume their work after they had left it for a period of time. Thus, they opened their shops to maintain the machines. It also prompted a relative increase in the work of merchants selling cloth and thread suppliers. "Upon buying the silk, the silk merchant could not believe it," said the Doumaniya. "When we provided the female workers who joined the project with machines, the lathe worker opened his shop and returned to work"[179]. She confirmed that many of the workers returned to work after a break, especially the female workers who left the industry for years. They continued to operate from 2012 until the city was besieged in late 2013, after which it was no longer possible to deliver raw materials to them or to receive products from them for marketing in Damascus or other countries.

After the siege ended, the initiatives moved to concentrate on Damascus, working with the displaced female workers to help them secure their daily livings. The women workers embroider at their homes, often avoiding going outside their residential areas for fear of being interrogated or arrested at security checkpoints. This led some bosses to start going from house to house to collect products for marketing, while others still receive products in their shops as before. The merchants, along with the NGOs, also helped the female workers to secure embroidery machines and raw materials, besides helping craftsmen to get loans that help them to secure what is necessary to practise their profession again, as was the case with the machine repairman, who confirmed his resuming his work after his displacement from Douma to Damascus. "Two years ago, when I came out of the siege and was able to overcome my personal problems, I resumed my work in my

---

[179] Lana Mradni (the author), "On the Aghabani Industry in Damascus", an interview with the Doumaniya (a nickname).

profession, the profession I love"[180], he said. The same is true for the merchant Abu Bassam, who expressed his love for his work and his desire to preserve this heritage despite the war circumstances, keeping hope for the return of past conditions[181].

The relationship between artisans and traders with the industry is not limited to its being part of their heritage or a mere profession, but also stems from their association with their craft since they learned it and it became part of their identity, for it is the space that enables them to create and innovate. They did not sell their machines despite their urgent need for money, instead they resumed their work when conditions improved and the demand grew, even if they had to work day by day. Despite the circumstances experienced by the Abu Saeed, the machine repairman, he confirms that "the war has made the industry abandon me rather than me abandoning it myself. My work has stopped completely for about two and a half years, but I have stuck to it"[182]. He received a loan from a local non-governmental organization to buy a lathe and work on his own.

The Aghabani became a consolation to some women after they lost one or more of their family members because of the war or detention; thus, it has become a source of hope. The merchants defied the war as they continued to manufacture and trade despite the lack of demand, refusing to change their trade, sell other products or travel, or in some cases they left the craft for a while before resuming it anew. "During the seven years of war, each one of us (who were working in this industry) found one way or another to meet our livings," said the printing worker. "I was one of them. However, once the opportunity came to resume my work again, I took out my machines that I kept, hid and did not sell, despite my need for their value, and resumed my work again, for a handicraft is a matter of addiction for a craftsman"[183]. The fact that some workers are attached to heritage and have the sense of

---

[180] Lana Mradni (the author), "On the Aghabani Industry in Damascus", an interview with Abu Saeed, the machine repairman (a nickname).
[181] Lana Mradni (the author), "On the Aghabani Industry in Damascus", an interview with the merchant Abu Bassam (a nickname).
[182] Lana Mradni (the author), "On the Aghabani Industry in Damascus", an interview with Abu Saeed, the machine repairman (a nickname).
[183] Lana Mradni (the author), "On the Aghabani Industry in Damascus", an interview with Abu Abdo, the printing worker (a nickname).

responsibility for their families' legacy is what motivates them to continue the industry in the hope that it will return to normal[184]. The labour force is still available, and alternatives to scarce raw materials can still be provided, benefiting from the opening of new marketing channels in foreign trade and some opportunities in internal markets.

The role played by the workers in this industry refers to its being a form of cultural, social and political resistance in the face of present-day violence[185] motivated by the desire to preserve their existence and identity[186]. This resistance seeks, according to Patricia Leavy, to confront the unrest through changes in jobs, systems and returns, considering this type of art a kind of alternative artistic practice. This type of art is not used directly to attack the dominant and despotic power, as in other forms of arts (such as theatre and the visual arts). Rather, it looks at the experiences of war and suggests a different history of individual narratives by the actors in handicrafts in order to provide another facet of a more humanistic war in which there are manifestations of resistance to death, the blurring of identities and the destruction of heritage, according to Dietrich Heissenbüttel[187]. This is represented in the insistence of the sample members on adhering to the industry and resuming their work, despite having quit it due to the siege and displacement, as it is the case with the machine repairman, the embroidery worker and printing worker. This comes from their insistence and desire to preserve the heritage and hope for the return of the industry to its previous status, as the merchants among the sample members said. As for the Doumaniya and her story with the Aghabani industry, she had experienced several stages since she learned it without her family's knowledge and relied on it to pay the tuitions of her university studies despite the opposition of her father, who deprived her of pocket money. Today, she is preoccupied with thinking of her two sons who were killed in the war[188].

---

[184] Lana Mradni (the author), "On the Aghabani Industry in Damascus", an interview with the merchant Abu Bassam (a nickname).
[185] Patricia Leavy, Method Meets Art: Arts-Based Research Practice, p. 216.
[186] Michaela Crimmin and Elizabeth Stanton, eds., *Art and Conflict*. London: Royal College of Art, 2014. Available online at https://bit.ly/2ZrikWS, p.10.
[187] Dietrich Heissenbüttel, "Art in Conflict: Interventions in War and Crisis", p. 4.
[188] Lana Mradni (the author), "On the Aghabani Industry in Damascus", an interview with the Doumaniya (a nickname).

The Aghabani industry was a source of living for many women who lost their breadwinners due to the consequences of their participation in the peaceful protest movement or to taking up arms along with the opposition factions, yet it was also a form of peaceful resistance during the protest movement in the city of Douma, and it might still be so for some women. This resistance represents a humanitarian challenge that people face to resist daily manifestations of death, according to Edward Said, who said that "humanity is the only and last form of resistance we have in the face of inhuman and unfair practices."

At the end of this part of the research, we can say that the war has crystallised the relationship of the research sample with the industry, since they established their association with identity and their resistance to the manifestations of death and displacement by continuing to work in a traditional industry inherited by some of them from their parents, learned since childhood or formed as a way out of obstacles caused by social pressure. Despite the fragmentation of the industrial community, the insistence of individuals to preserve it is a key factor in preserving it from extinction, and concerted efforts of all concerned might be useful to achieve the optimal conditions that allow the development of the industry. The research suggests that all the above-mentioned reactions of workers toward the Aghabani industry should be considered under the rubrics of cultural resistance and resilience through a kind of traditional arts, through which they create their own narratives, which we recommend to expand the understanding of the reality of the industry and the reality of the research community in Douma and Damascus in particular.

Conclusion

This study attempted to contribute to monitoring the change in the motives of individuals or groups active in the Aghabani industry, and the resulting actions that challenge the effects of the war and its risks on the industry, by asking about the motivation for workers in the industry to find solutions and mechanisms that will contribute to the craft's continuation and preservation from extinction, based on this industry's heritage value and its being a reflection of their cultural identity and a basis for their cultural resistance to the circumstances and challenges of the war.

The research was based on qualitative methodology, which contributed to highlighting the social and dynamic context of the Aghabani industry before 2011 and up until the present time—that is, the end of 2018—through semi-structured non-random in-depth interviews with a sample of members selected from the research community, i.e. active actors in the Aghabani industry.

First of all, the research reached a collection of documentary information concerning the establishment of the industry dating back to the early Ottoman era, when the Aghabani turban was a certificate of the craftsman's mastery[189], recognised in a craftsmanship system brought in by Sultan Suleiman the Magnificent in the sixteenth century[190]. While the research initially assumed that the Aghabani industry was adversely affected by the war and therefore at risk of extinction, it turned out that the industry has rebounded because the efforts of Doumani female workers who returned to work to earn their livings after losing their husbands and sons in the midst of events.

The research also discussed the industry's state prior to 2011 and discovered a number of factors that contributed in one way or another to undermining the potential of the industry. These included issues related to phenomena like random urbanization and the decrease in the number of female workers because of the higher rate of education for girls, in addition to government policies that neglected heritage

---

[189] Najat Qassab Hassan, Damascene Discourse: Memoirs 1, 1884–1983, p. 260.
[190] Nasaar Abdul-Azim, "Craftsmanship Organization and Local Industries in the Ottoman Era".

industries in general before 2011. After 2011, however, governmental and non-governmental organizations such as the General Union of Craftsmen and the Syria Trust for Development carried out a series of measures, such as training artisans, buying products for marketing and selling in the markets of friendly countries like Russia and China, and securing handicraft areas where displaced artisans can work[191]. However, it is hard to answer the question regarding the extent to which these measures affect the process of spontaneous development of the artisanal industries in general and the Aghabani craft in particular, which is considered a danger that affects the dynamic of the development of the industry. UNESCO did not object to that; on the contrary, it praised the role of the Syrian government and the Syria Trust for Development[192] in protecting the Syrian intangible heritage during a time when the military machine, in which the Syrian government is the main actor, is contributing to the destruction of the environment of the industrial society and the death and displacement of its members in a way that would hinder the transmission of the intangible heritage from one generation to another[193]. This raises a number of questions that should be addressed; questions related to the mechanism of UNESCO's work and the criteria for its evaluation of the performance of states and their institutions in the preservation of the intangible cultural heritage and its ability to apply them away from political considerations.

The research found that individuals' motivation and adherence to the industry are related to their sense of identity and form a kind of cultural resistance[194] to the challenges of war, death and displacement. It also pointed out that this identity link was a generator of the cultural resilience[195] that compels them to adapt to the changes to ensure the

---

[191] "The Union of Craftsmen Adopted a Brand for Globally Desired Crafts".
[192] UNESCO, "Report by a Non-Governmental Organization Accredited to Act in an Advisory Capacity to the Committee on Its Contribution to the Implementation of the Convention".
[193] "Syria: UNESCO Establishes an Observatory for the Safeguarding of Cultural Heritage".
[194] Patricia Leavy, *Method Meets Art: Arts-Based Research Practice*, p. 216. See also Jocelyn Hollander and Rachel Einwohner, "Conceptualising Resistance", pp. 533–554.
[195] Stephanie J. Rotarangi and Janet Stephenson, "Resilience Pivots: Stability and Identity in a Social-Ecological-Cultural System".

preservation of their intangible cultural heritage and its transmission to future generations, considering the Aghabani industry a kind of art that contributes to the creation of personal narratives that provide other readings of the war[196] and part of human practice in the face of the inhuman practices among which they live[197]. However, there is a need for the concerned parties to intervene to ensure that the community of people involved in the Aghabani industry is involved in how to develop the industry to adapt to the current circumstances[198] without external interference that may lead to a change in this society's priorities.

Therefore, the research proposes the following set of recommendations as steps that must be taken to help the Aghabani industry to succeed in its mission to preserve its intangible cultural heritage:

- Follow up on the research and documentation of information on the history of this craftsmanship and conduct in-depth study of the reasons for the decline in industry before 2011 and the subsequent working conditions, whether it relates to the industry incubators or the relevant authorities of the Syrian government such as the Syria Trust for Development or other non-governmental institutions.
- Provide a database upon which cultural policies of the Aghabani industry and heritage industries in general can rely in the near and long term. There are no sufficient studies on the industry, its history or development.
- Document the stories of the people working in the industry to be part of the Syrian cultural memory, a vivid example of cultural resistance, and forms of people's challenge to the siege and death through their persistence to keep working in the Aghabani industry in particular and all handicrafts in general.
- Investigate the mechanism of UNESCO's work with governments and its institutions and exert pressure through

---

[196] Dietrich Heissenbüttel, "Art in Conflict: Interventions in War and Crisis", p. 4.
[197] Edward Said, *Humanism and Democratic Criticism*, 1st ed., Columbia University Press, New York, 2004, p. 26.
[198] Graciela Singer, "The Importance of the Intangible Cultural Heritage".

institutions or cultural initiatives to act as effective monitors over the mechanism of the Syrian government, all concerned bodies and other non-governmental institutions to find ways to restore the incubating environment of the industry in its home region, especially the city of Douma, and ensure the transmission of the industry's knowledge and skills in a spontaneous process without external interference.

- Press UNESCO and the Syrian government to develop working standards to safeguard the reputation of the handicraft industry's products in general, and the Aghabani in particular, by registering the industry; protecting it from competition with imported products, and from theft and reproduction by other countries; and by cooperating with artisans to find solutions to the industry's current problems, like the lack of raw materials, the need to re-open the state factories for the manufacture of cloth, the displacement of female workers and craftsmen, and the issue of marketing the products internally and externally. The government should also provide consumer protections by establishing logical controls calculating the cost price of local products applied to everyone, taking into account the war conditions and its various effects.
- Develop a monitoring mechanism to ensure that artisans and the industry community benefit primarily from marketing revenues, protect their efforts from exploitation and ensure their participation in the process of industrial development and modernization of products.
- Encourage young women and men to learn identity-related crafts through the re-development and upgrading of the "sheikh of crafts" system for quality assurance and high-quality trade transmission based on exams, in addition to benefiting from modern science in developing some crafts that enable the development of mechanisms and maintain the aesthetics of handmade products, like repairing embroidery machines.
- Support civil initiatives aimed at supporting the industry's incubating community.

In the end, this study confirms the need to expand future research, to examine the roles of both the Syrian government and its institutions such as the General Union of Craftsmen and the Union of Craft Associations and the affiliated institutions such as the Syria Trust for Development and the Syrian Handicrafts company, in addition to the role of international organizations concerned in safeguarding the intangible cultural heritage in general and the Aghabani industry in particular. Further research should also follow up the fate of workers and craftsmen displaced from Douma after the city's return to the Syrian government's control and the dynamism of the development of the industry and the challenges resulting from it and the mechanisms of resistance developed by workers in both Douma and Damascus.

References

Books

Al-Allaf, Ahmad. *Damascus at the Beginning of the Twentieth Century*, translated by Ali Naisse. Damascus: Syrian Ministry of Culture, 1976.

Fayyad al-Fayyad, Mohammed, and Majed Hashim Hammoud. *Traditional Crafts in Syria*, first edition, translated by Majd Hamoud. Damascus: General Union of Craftsmen, Office of Culture and Media, 2011.

Kayyal, Mounir. *Levantine Achievements in Damascene Arts and Industries*. Damascus: Syrian Ministry of Culture, 2006–2007.

Kayyal, Mounir. *Damascene Arts and Industries*. Damascus: Syrian Ministry of Culture, 1985.

Maallah, Talal. *The Non-Material Heritage: Skills Associated with Traditional Craftsmanship, Part I*. Damascus: Syrian Ministry of Culture and Syria Trust for Development, 2014.

Qassab Hassan, Najat. *Damascene Discourse: Memoirs 1, 1884–1983*. Damascus: Tlass Publishing House, 1988.

Interviews

Mradni, Lana (the author). "On the Aghabani Industry in Damascus", an interview with the Doumaniya (a nickname), (n.p.), Damascus, April 28, 2018.

Mradni, Lana (the author). "On the Aghabani Industry in Damascus", an interview with the merchant Abu Bassam (a nickname) (n.p.), Damascus, on March 3, 2018.

Mradni, Lana (the author). "On the Aghabani Industry in Damascus", an interview with the merchant Abu Samer (a nickname) (n.p.), Damascus, on April 21, 2018.

Mradni, Lana (the author). "On the Aghabani Industry in Damascus", an interview with the Abu Abdo, the printing worker, (a nickname), (n.p.), Damascus, on March 23, 2018.

Mradni, Lana (the author). "On the Aghabani Industry in Damascus", an interview with Abu Saeed, the machine repairman, (a nickname), (n.p.), Damascus, on April 23, 2018.

Mradni, Lana (the author). "On the Aghabani Industry in Damascus", an interview with Abu Muhammad, the embroidery worker (a nickname), (n.p.), Damascus, on April 23, 2018.

Academic Studies

Abdul-Azim, Nasaar. "Craftsmanship Organization and Local Industries in the Ottoman Era" [electronic reference]. University of Kufa, *Journal of the Kufa Center for Studies*, Issue 43, available at https://bit.ly/2HYCiCG, published in 2016, accessed on April 4, 2019.

Masri, Adnan. "The Impact of Urbanization in Damascus on Passenger Transport Movement: A Ph.D. Thesis in Economic Geography" [electronic reference]. Damascus University, Faculty of Arts and Humanities, Geography Department, 2014–2015, available at https://goo.gl/M2FPwc.

Al-Nahar, Ammar. "A Study on the Book of Traditional Crafts in Syria". *Journal of Creative Circles*, Third Issue, Damascus University, 2015.

Official Publications

Syrian Arab Republic: Syrian Prime Ministry, Law No. 14 of 2015 on Internal Trade and Consumer Protection [electronic reference]. Damascus: Prime Ministry of the Syrian Arab Republic. Available at https://goo.gl/u6B4uq, accessed on April 3, 2019.

UNESCO. "Convention for the Safeguarding of the Intangible Cultural Heritage, Article 2: Definitions". Adopted in Paris on October 17, 2003. Available on the UNESCO Intangible Cultural Heritage website at https://bit.ly/2OKUobK.

UNESCO. "Convention for the Protection of Cultural Property in the Event of Armed Conflict" ("The Hague Convention"). Adopted in the Hague, Netherlands, on May 14, 1954. Arabic text available on the website of the University of Minnesota Human Rights Library at https://bit.ly/2GnnMRZ, accessed on May 23, 2018.

Articles

Dandashli, Mustafa. "Dr. Abdul-Karim Rafik: The Manifestations of Craftsmanship Organization in the Levant in the Ottoman Era" [electronic reference]. Cultural Centre for Research and Documentation at Sidon, Lebanon. Available at https://bit.ly/2v4gq0i, published on April 1, 1981, accessed on 15/09/2018.

Arab48. "A Mass Exodus from Douma and Thousands of People Leaving Ghouta for Idlib" [electronic reference]. On the *Arab48* news website. Available at https://goo.gl/RTF9tL, published on March 26, 2018, accessed on May 23, 2018.

Al-Hajali, Bashar, et al. "Traditional Crafts and Industries: A Homeland's Memory. How Can We Restore Its Spark?" [electronic reference]. *Al-Thawra Online*. Available at https://goo.gl/7nPGqx, published on December 22, 2012, accessed on September 7, 2018.

Balanche, Fabrice. "A City under the Control of the Baath: Damascus, the Syrian Capital" [electronic reference]. *Al-Jumhuriya* website. Available at https://goo.gl/RWpWf9, published on June 14, 2012, accessed on May 27, 2018.

Al-Asaad, Omar. "Creative Forms in the Syrian Revolution: The Coordinating Councils as a Model" [electronic reference]. *BabelMed*. Available at https://goo.gl/jyYqEz, published on June 20, 2012, accessed on January 25, 2019.

Kakerli, Samer. "With Figures, Orient Net Investigates How Assad Destroyed Damascus' Ghouta" [electronic reference]. *Orient Net*. Available at https://goo.gl/AQuqtj, published on September 8, 2014 accessed on June 6, 2018.

Zeno, Basileus. "Neither the People nor the Monuments: Palmyra and the War Tragedies of the Ruins" [electronic reference]. *Al-Akhbar* newspaper. Available at https://goo.gl/ffcyJn, published on June 1, 2015, accessed on June 3, 2018.

Afra, Mohammed. "Syria's Student Dropout Problem: A Phenomenon Exacerbated by Poverty, Customs and Lack of Law Enforcement" [electronic reference]. Deutsche Welle's DW Akademie. Available at https://goo.gl/tuZn1t, published on February 21, 2010, accessed on June 7, 2018.

"Douma Protests on 'No to Federalism' Friday" [electronic reference]. *Syria Untold*. Available at https://goo.gl/wRhMXA, published on April 1, 2016, accessed on 3/5/2018.

Saleh, Rasha. "Syrian Women: Suffering and Challenges" [electronic reference]. *Al-Nabad*. Available at https://goo.gl/kyNxdA, published on April 19, 2017, accessed on May 26, 2018.

Ghosn, Ziad. "Thus the Doumani Women Revived Syria's Heritage" [electronic reference]. *Al-Akhbar*. Available at https://goo.gl/s8EFaM, published on May 22, 2017, accessed on May 23, 2018.

Telfah, Fadi. "The Encyclopedia of Traditional Crafts: A Documentation Aimed at Protection" [electronic reference]. eSyria. Available at https://goo.gl/BtQdjj, published on June 11, 2012, accessed on September 7, 2018.

Hanawi, Maada. "With Its Workers' Expertise and Its Own Efforts: The General Maghazel Company Resumes Production" [electronic reference]. *Tishreen*. Available at https://goo.gl/pQ41zb, published on October 27, 2016, accessed on June 7, 2018.

Jabbour, Sulafa. "Damascus' Eastern Ghouta: A Hundred Days under Siege" [electronic reference]. *Al Jazeera*. Available at https://goo.gl/gcphsR, published on January 12, 2014, accessed on June 1, 2018.

Al-Mustafa, Bassam. "Our Traditional Craft Industry and the Dangers of Its Counterfeit?!" [electronic reference]. *Al Azmenah*. Available at https://goo.gl/ATjGpg, published on December 20, 2012, accessed on August 29, 2018.

"A Recruitment Campaign for the Remaining Young Syrians Forcing Them to Emigrate" [electronic reference]. *Asharq Al-Awsat*. Available at https://goo.gl/fcQA4A, published on December 21, 2016, accessed on May 3, 2018.

Jani, Yasmin. "The Status of Ancient Arab Houses to the Syrians" [electronic reference]. *Zaiton*. Available at http://www.bit.ly/2UjfPHa, published on July 22, 2017, accessed on August 12, 2018.

SANA. "A Workshop on the Reality of Traditional Industries" [electronic reference]. SANA (Syrian Arab News Agency). Available at https://goo.gl/yuBc31, published on August 17, 2016, accessed on May 3, 2018.

"Ubbaha Revives Syrian Handicrafts in al-Takiyya al-Sulaymaniyah" [electronic reference]. *Al-Boselh*. Available at https://goo.gl/TfKcfB, published on October 22, 2015, accessed on June 7, 2018. See also Ubbaha's official Facebook page at http://bit.ly/2FLynX3.

"Activation of 14 Crafts Benefit 142 Craftsmen with Revenues of 84 Million" [electronic reference]. *Al-Baath*. Available at https://goo.gl/1vQXem, published on August 29, 2016, accessed on May 10, 2018.

Diab, Abdul Razzak. "al-Takiyya al-Sulaymaniyah in Damascus" [electronic reference]. *The New Arab (Al-Araby Al-Jadeed)*. Available at https://goo.gl/Zb737V, published on February 27, 2018, accessed on September 7, 2018.

"Douma, a New Chapter of Forced Displacement in Syria" [electronic reference]. Sky News. Available at https://bit.ly/2OSyYcN, published on April 9, 2018, accessed on September 15, 2018.

"The Union of Craftsmen Adopted a Brand for Globally Desired Crafts" [electronic reference], *Al-Thawra Online*. Available at https://bit.ly/2KLQtgB, published on December 12, 2016, accessed on September 7, 2018.

UNESCO. "The Intangible Cultural Heritage" [electronic reference]. On the website of UNESCO's Cairo Office. Available at https://goo.gl/abeafB, (n.p.), accessed on April 3, 2019.

UNESCO. "The Urgent Safeguarding of the Syrian Intangible Cultural Heritage" [electronic reference]. UNESCO Syrian Cultural Heritage Observatory. Available at https://goo.gl/HjdcUH, (n.p.), accessed on April 6, 2019.

UNESCO. "'Stop the Destruction of Syrian Cultural Heritage!' Urges UNESCO Director-General" [electronic reference]. Available at http://bit.ly/2JhDNdg, published on August 29, 2013, accessed on April 4, 2019.

UNESCO. "The Intangible Heritage" [electronic reference]. UNESCO Syrian Cultural Heritage Observatory. Available at https://goo.gl/3b8Uog, (n.p.), accessed on April 4, 2019.

United Nations. "Syria: UNESCO Establishes an Observatory for the Safeguarding of Cultural Heritage" [electronic reference]. UN news site. Available at https://goo.gl/yBPT5e, published on May 29, 2014, accessed on April 3, 2019.

Mansour, Karam. "The Syria Trust for Development: From the Civil Face to the Military Uniform" [electronic reference]. *Al-Jumhuriya*. Available at https://goo.gl/ESWfwM, published on August 14, 2017, accessed on June 25, 2018.

"The Government, in Cooperation with the Syria Trust for Development, Has Allocated 10 Billion Syrian Pounds to the Project of the Homeland's Wounded" [electronic reference]. Syria Steps. Available at https://bit.ly/2WWuhBG, published on April 13, 2017, accessed on June 25, 2018.

"The Syria Trust for Development Has Launched the Project of the Laurel Wreath" [electronic reference]. Tishreen University. Available at https://goo.gl/K65chx, (n.p.), accessed on May 3, 2018.

Abdo, Batoul. "Syrian Exports: Crisis Scenarios and Implications" [electronic reference]. *Al-Thawra Online*. Available at https://goo.gl/iYpmBE, published on March 20, 2012, accessed on June 6, 2018.

Adra, Hisham. "Syria's Handicrafts and Traditional Knowledge Are at Risk of Loss" [electronic reference]. *Asharq al-Awsat*. Available at https://goo.gl/F95Lgj, published on February 3, 2015, accessed on June 28, 2018.

"On World Environment Day, the Urban Expansion of Damascus' Ghouta Claims 4,000 Hectares of Agricultural Land" [electronic reference]. *Tishreen*. Available at https://goo.gl/ov73MN, published on June 7, 2006, accessed on May 10, 2018.

Al Jazeera Encyclopedia. "Douma: Ghouta's Bride Targeted by Blockade and Poisonous Gases" [electronic reference]. Available at https://bit.ly/2CSXqpN, published on April 7, 2018, accessed on January 24, 2019.

Al Jazeera Encyclopedia. "The Eastern Ghouta: A Syrian Region under Destruction" [electronic reference]. Available at https://goo.gl/Er45Bu, published on February 21, 2018, accessed on August 2, 2018.

Al-Sa'o, Sabreen. "The City of Douma in Syria" [electronic reference]. Available at https://goo.gl/sgYsTM, published on April 4, 2017, accessed on August 12, 2018.

Al-Najem, Ismail. "Mounir Kayyal: A Documented Encyclopedia as a Tribute to the City: 20 Books Telling the Details of the Levant" [electronic reference]. eSyria. Available at https://goo.gl/HVGcKh, published on November 20, 2014, accessed on May 23, 2018.

Kholouf, Hassan. "The Surroundings of Damascus and Urban Havoc" [electronic reference]. Syria News. Available at https://goo.gl/V2EV9L, published on January 6, 2011, accessed on September 7, 2018.

Safdie, Hossam. "The Expansion of Damascus between the Dreams of Strategic Vision, the Tragedy of the Nearby Countryside and the Reality of a Japanese Study" [electronic reference]. Rational and Guided Municipality Forum. Available at https://goo.gl/g8v16V, published on February 27, 2011, accessed on July 15, 2018.

Videos

Al-Assad, Asma. Speech to employees of the Syria Trust for Development. On YouTube. Available at https://goo.gl/83cb7Y, published on July 22, 2017, accessed on May 3, 2018.

"Fire Braids: The Women who Fought with the Government Army". A video featuring comments by first lady Asma al-Assad. On YouTube. Available at https://goo.gl/5D86y, published on March 22, 2018, accessed on May 3, 2018.

References in English

UNESCO. "Convention for the Safeguarding of the Intangible Cultural Heritage, Article 2: Definitions". Adopted in Paris on October 17, 2003. UNESCO Intangible Cultural Heritage website. Available at https://bit.ly/2OKUobK, accessed on April 3, 2019.

UNESCO. "Report by a Non-Governmental Organization Accredited to Act in an Advisory Capacity to the Committee on Its Contribution to the Implementation of the Convention" [electronic reference]. UNESCO Intangible Cultural Heritage website. Available at https://goo.gl/EBkaas, published on January 18, 2017.

Said, Edward. *Humanism and Democratic Criticism*, 1st ed. New York, NY: Columbia University Press, 2004.

Ahamad, Fayaz, and Effat Yasmin. "Impact of Turmoil on the Handicraft Sector of Jammu and Kashmir: An Economic Analysis". Srinagar, India: Department of Economics, University of Kashmir. *International NGO Journal*, Vol. 7(5), 2012, pp. 78-83.

Mack, Natasha, and Cynthia Woodsong, Kathleen M. Macqueen, Greg Guest, and Emily Namey. *Qualitative Research Methods: A Data Collector's Field Guide*. Research Triangle Park, North Carolina: Family Health International, 2005.

Singer, Graciela. "The Importance of the Intangible Cultural Heritage" [electronic reference], on *Terrae Antiqvae* (blog), available at http://bit.ly/2XH736y, published January 5, 2011.

Lenzerini, Federico. "Intangible Cultural Heritage: The Living Culture of Peoples". *The European Journal of International Law*, Vol. 22, No. 1, February 2011, pp. 101–120.

Halbwachs, Maurice. *On Collective Memory*. Trans. and ed. Lewis A. Coser. Chicago, Illinois: University of Chicago Press, 1992.

De Cesari, Chiara, & Ann Rigney (eds.). *Transnational Memory: Circulation, Articulation, Scales*. Berlin/Boston: Walter de Gruyter GmbH, 2014.

Hollander, Jocelyn A., & Rachel L. Einwohner. "Conceptualizing Resistance". *Sociological Forum*, Vol. 19, No. 4, December 1, 2004, pp. 533-544.

Sweeney, Simon. "Review: Reconstructing Spain: Cultural Heritage and Memory after Civil War". *International Journal of Heritage Studies*, Vol. 17, No. 6 (November 2011), pp. 629–631.

Rotarangi, Stephanie J., and Janet Stephenson. "Resilience Pivots: Stability and Identity in a Social-Ecological-Cultural System". *Ecology and Society*, Vol. 19, No. 1 (March 2014), available at https://goo.gl/uWek8P.

Leavy, Patricia. *Method Meets Art: Arts-Based Research Practice*. New York, NY: Guilford Publications, 2015.

Heissenbüttel, Dietrich. "Art in Conflict: Interventions in War and Crisis". An essay published in the context of the Global Activism exhibition in Karlsruhe, Germany, 2014.

Kalter, Johannes, and Margareta Pavaloi, Maria Zerrnickel and P. Behnstedt. *The Arts and Crafts of Syria*. London: Thames and Hudson, 1992.

Crimmin, Michaela, and Elizabeth Stanton, eds. *Art and Conflict*. London: Royal College of Art, 2014. Available online at https://bit.ly/2ZrikWS.

Healey, Susan. "Cultural Resilience, Identity and the Restructuring of Political Power in Bolivia". A paper submitted for the 11th biennial conference of the International Association for the Study of Common Property, Bali, Indonesia, June 19–23, 2006.

Appendix

A sample of the structure of the interviews conducted with members of the research sample

Personal information:

- Name
- Age
- What is your job within this business/enterprise?
- When did you start working in the Aghabani industry?

Research-Related Questions:

1. Tell us about the Aghabani industry, its history, place of origin and development.
2. Why do you work in the Aghabani? What does this industry mean to you?
3. For how many generations has your family worked in this industry?
4. Among your family members, who works in the Aghabani industry today?
5. What are the motivations that made you continue this work? How did the war affect and change your business? How was that?
6. How did the war affect the course of your job in terms of the course of the production process?

   => Labour availability/labour migration
   => Availability of raw materials

=> Marketing and selling products in the domestic and foreign markets

7. How did the war affect the course of your job in terms of the marketing process?

8. Which countries did the work reach before the war, and which countries does it reach today?

9. What are the solutions that have been found regarding the problems related to:

    => Labour force
    => Raw materials
    => Marketing & Sales

10. Could you please identify some of the negative and positive impacts of the war circumstances on the Aghabani industry?

11. If you believe that some of the industry's production and marketing areas have been developed as a result of thinking about alternative methods, please explain.

12. What are the risks that have not been solved yet, and that are facing the industry at the moment?

13. What are the strengths of the project/trade?

14. What are the weaknesses of the project/trade?

15. Which is the most influential actor or element on this industry:

    => Governmental institutions
    => The market
    => The security situation
    => Civil society

16. What recommendations and proposals do you suggest to cultural institutions concerned with heritage industries?

17. Do you think the Aghabani industry exists or previously existed in a city other than Douma?

18. Why do you think that the city of Douma alone has specialised in this industry?

19. From where do you usually buy the following raw materials?

The Material	Before 2011	After 2011
The embroidery threads		
The cloth		
The printing materials		

20. Could you please explain the work dynamic before 2011?

21. Could you please explain the work dynamic after 2011?

Dima Nachawi

Born in Damascus in 1980, Nachawi holds a Master's degree in Art and Culture from King's College in London in 2017. She is a graduate of Damascus University's Faculty of Humanities, Social Science Studies Department in 2003. She worked in the field of animation, graphics, relief and social services for war-affected people at UNHCR Damascus.

She currently lives and works in Beirut as a graphic artist, animator, narrator, and clown. She is the founder of Musk Project, which is an initiative to transform current Syrian memory into illustrated stories. She is a member of the Clown Me In group, which aims to alleviate the suffering of refugees and raise awareness about a range of local social issues, where she works in the management of one of its projects titled "Forced Jesting."

Changes in the Gender Roles of Women Handicraft Workers in Damascus After 2011

An Exploratory Paper

By: Ola Alshikh Hassan

Under the supervision of Hassan Abbas, PhD

Summary

Since the beginning of the twentieth century, wars have played a role in transforming gender roles, with women starting to practice jobs that were previously considered exclusively for men. This happened during the Second World War because of the shortage in men. Similarly, as a result of the continuing civil war in Syria, a large number of male workers have been lost since 2012 and women who were dependent on male breadwinners have been forced to work to meet their needs.

From this point of view, this paper raises a number of questions about the gender role changes experienced by women in Damascus and its countryside under such exceptional conditions. Will women continue to work if their previous life conditions return? Will they stop working? Will the accompanying changes stay, or will women return to the more stereotyped roles they played in the past? Can these changes these working women have experienced turn them into agents of change in their small communities? The importance of this study comes from trying to monitor the nature of these women's entry into the labour market, whether it is motivated only by need, and if such induced changes will end once the job is not needed anymore.

In order to explore this change, its conditions and components, this research relied on a sample of women working in handicrafts at home for companies, individual customers or market merchants. Thirty-five women between the ages of 25 and 55, most of whom were from the Ghouta region of Damascus and are currently residing in relatively safe places in Damascus and its countryside, were selected for the study. Most of the women are in full charge for their children. They have been displaced from the Ghouta region and have lost their sources of income and assistance, such as the husband and sometimes their brothers or husband's families. This research followed the women over a three-year period between 2015 and 2018.

Key findings of the research are that economic and social conditions have led women to work and that the absence of the husband has shifted the head of the hierarchy in the family to the mother, while displacement has facilitated changes in dress and necessitated other changes in mobility and self-reliance in the conduct of personal and family affairs. Women also confronted a radical change in the thinking system of their communities and became breadwinners for themselves and their families, gaining a rapid development in the sense of their ability to act in general, and forming a strong basis for taking charge of the personal decisions that previously were interfered with by other family members. Women have overcome the problem of being able to interact with other societal categories in the context of what their work entails. While the change in their stereotypical roles has not led to a direct or significant change in the habits and values that regulate gender relations, it has enriched their power and ability to make decisions for themselves and their children.

Introduction

After the outbreak of civil war in Syria in 2011, many Syrian women resorted to work in handicrafts at their homes as a result of displacement, the deterioration of the economic situation, and the absence of a male breadwinner in many households. These women suddenly had to take responsibility to support their families and children. The choice of working at home was suitable for several reasons. First, this provided the women with the possibility of staying at home with the children and taking care of them, especially in cases of displacement to a new residence place and living away from their usual environment. It also allowed them to avoid breaching the customs and traditions of conservative communities that do not encourage the exit of women outside the home, even if it is for work. Working at home also allowed the women's parents to have control over them in cases of temporary or permanent absence of the husband. In addition, this form of work is familiar and accepted, and not a new one, among the more conservative Syrian social classes that do not mind the idea of women's work. Rather, these communities focus on rejecting women's mixing with men or leaving their homes.

Since the beginning of the twentieth century, wars have played a role in overcoming gender stereotypes, when women would start to work in jobs once considered to be the exclusive domain of men. During the Second World War, due to a shortage of males, European women gained additional rights and entered new production fields in industry and agriculture[1]. Likewise, the war in Syria has caused a large decline in the number of working males since 2012, and women, who used to depend on male breadwinners, have become forced to work to meet their families' needs.

At first, Syrian women entered the labour market only partially and in specific sectors, and the jobs they performed were not exactly similar to those of women in Western countries, often due to social and religious barriers. However, this led to changes in ways of thinking

---

[1] National WWI Museum and Memorial, "Women in World War I" [electronic reference], available at https://bit.ly/2Kyis3h, (n.d.), accessed on February 10, 2018.

about gender that affected Syrian women's lives. This need-driven entry into the labour market raises questions as to whether these changes are permanent, or will they end as the conditions that made them necessary fade away? Will women continue to work if the conditions of their former lives are restored to how they were before? Will they stop working? Will the associated changes in women's gender roles be confirmed in that case, or will women revert to the more stereotypical roles? Can these changes lead to women's becoming change agents in their small society?

This research explores those questions and attempts to list and explain the changes in the gender roles of women who have entered the labour market in Damascus and its countryside since 2011. It questions what these changes are and how they actually affected the lives of women involved in the labour market by comparing the current situation to how it was before 2011, when most of these female workers' families were displaced and forced to leave their homes. The effects of this displacement included disruptions in the basic requirements of life, the loss or absence of the husband in most cases, and being separated from the rest of the larger extended family as a result of dispersion and the various situations of relatives.

In order to explore this change, its circumstances and components, this research relied on a sample of women who do handicrafts work at home for companies, individual clients or traders. Thirty-five women, aged 25 to 55, mostly from the Ghouta region of Damascus' countryside, were selected. They now live in safe areas in Damascus and its countryside, where they were interviewed in the first half of 2018. Most of them are fully responsible for raising their children. Many were displaced from the Ghouta area, and lost their sources of income and assistance provided by husbands and sometimes by brothers and the husband's family. Those who have not been displaced have still experienced negative changes in their financial conditions due to economic deterioration, the loss of the family's breadwinner or the changing work conditions as a result of the war.

The study relied heavily on the observations of the researcher, who works in a company that deals with the women. She followed them closely during a period of three years between 2015 and 2018. In addition, quantitative questionnaires were distributed to 30 women, focusing on the criteria governing gender-role changes in Syrian society. An extensive small-group meeting of six women was conducted, in which the differences the women saw in their lives were discussed from their point of view. Four in-depth interviews were conducted with four women workers from various regions, ranging in age from 30 to 50, who represent the largest proportion of the women surveyed working in handicrafts in Damascus and its countryside.

The questions were divided into two parts[2], before and after 2011. Whether the women worked before 2011 and the reasons for it are a factor influencing the study as a whole, denying or confirming the occurrence of change. Other factors surrounding the war and its consequences also play a role in the process of change in women's roles and their transition to economic independence, in comparison to their form of life before the war. These factors influence one another and add an impact varying in size from one experience to another. However, in the end they are the factors that led women to take control of their lives and adopt economic choice options.

Other questions dealt with the importance of intellectual change related to the concept of gender and stereotypical roles, represented by the women's satisfaction with their choice, a sense of its importance and an awareness of the achieved gains, and monitoring their way of thinking about the closest related persons, like the children.

The women of the sample who entered the labour market during this period belonged to the middle and poor classes before 2011. The majority of them lived in their own houses or at least they owned the contents and furniture of the houses. Living in proximity to the other extended family members was common, in accordance with the simple rural lifestyle in which the larger family contributed to the needs of living and caring for the children and relied on the presence of the husband or another male family member and their work as a source of income. This society focuses on preserving the stereotypical gender roles in accordance with the conservative social and religious customs in Damascus and its countryside. These roles are represented by considering women primarily as wives and mothers whose identity is limited to their reproductive role. They are prepared from the earliest age

---

[2] See the Appendix for the questionnaire used in this research and the questions asked during the interviews conducted with members of the study sample.

to do the necessary household chores of cleaning, cooking and caring for children, besides learning other skills such as sewing and embroidery.

The study takes into account the quality of the work practised by women. The transformation of handicrafts into an income source is a relatively new phenomenon for the women in the study sample and their former social environment, which has not been previously monitored due to the current state of instability in Syrian society and the relatively minor impact of handicrafts jobs in the home, compared to occupations that require greater commitment to regular working hours outside the home. However, the prevalence of handicrafts work among a class that is the most representative of the stereotypical roles of women in Syrian society, and its noticeable impact on the limited effectiveness of its practice in the public sphere, is worth highlighting.

Methodologically, this research is based, especially in the way of dealing with the sample of working women, on two books. The first is Najla Hamadeh's *Lentils and Caviar: Biographies of Women Who Live in Lebanon*[3], which reviews the life experiences of women living in Lebanon from the different social classes, representing a large sample of society through interviews conducted by the writer in Beirut. The second is *The Weight of the World*[4], by the French sociologist Pierre Bourdieu, in which he presents combined testimonies of men and women from around the world, accompanied by theoretical commentary, questioning complex social issues in a large urbanised country like France. Bourdieu brings together the faces of alienation in the capitalist society, relying on statements that reveal racism, violence and identity, yet he believes, at the same time, in the liberated self—that is, a person who makes connections between phenomena and their causes and believes that he or she is able to confront old phenomena and generate new ones.

This study is divided into two parts. The first focuses on the social and economic background of the working women surveyed and re-

---

[3] Najla Hamadeh, Lentils and Caviar: Biographies of Women Who Live in Lebanon: Self-Construct, Agency, and Independence, Beirut: Arabic Scientific Publishers, 2015.

[4] Pierre Bourdieu, *The Weight of the World: Social Suffering in Contemporary Society* (Arabic translation), Damascus: Kanaan Publishing House, 2001.

views the practical considerations that led to their going to work and influenced their choice of handicrafts. This section also examines the aspects that surround the lives of the women and social and economic changes, and the impact of those changes on the women, within the context of the war in the study area: Damascus and its countryside in particular.

In the second part, we review changes in gender roles related to work in particular and its financial and social effects, represented by women's gaining economic independence and taking financial responsibility for raising children, two aspects that were previously held by males in the family. These changes are directly connected to work, such as going out to secure the personal, family and material needs that are part of the life situation in which they are required to perform duties that concern them and their families. As a result of this situation, responsibility becomes inevitable due to the absence of a male. The ways women exercise these responsibilities give indications of how they think about themselves and the world, and how they have taken over the responsibilities they chose when they went to work. This part of the study also attempts to review the changes that have occurred in women's lives, how they perceive this form of change, and whether they desired it, since the challenge of their work is still imposed on them by their economic conditions rather than rising from their own will to seek it.

The main difficulties we encountered during our research were to differentiate between the effects of work on women as a source of strength and responsibility, and the impact of other circumstances, like the loss of men in the family, who are usually responsible for children and decision-making, and the women's relocation into a relatively strange society and surrounding environment that affects them through a change in their form of attire and greater contact with the male members of the family because of the crowded housing conditions forced on them due to higher rents. This made it difficult to determine the impact of work as one variable among other variables that had accompanied it or were the reason for their resorting to it.

Other difficulties included the study's aim to compare conditions between two periods, before and after 2011. The women's responses for the first period, before 2011, are subject to their current perceptions of the past, and this makes it difficult to monitor actual changes in their views or attitudes.

In order to monitor a change in stereotyped roles, the form of these roles must first be identified. The right of women to make decisions for themselves, for example, is an umbrella category that encompasses freedom of dress, freedom movement, choice of residence, choice of work and all that follows. This constitutes the core of the typical difference between males and females in the society to which the women belong. When changes took place in the context of these freedoms, they caused one another in succession. Work leads to the freedom and independence of economic decision. This independence of economic decision, however, was liberated by the transfer of responsibility in the management of the family to women after the absence of the male breadwinner. The absence of the male breadwinner was a major cause of action, and so on. This makes the untangling the intertwined cause-result relationships of these freedoms one of the most important challenges we faced during our work.

Finally, we would like to thank the working women who participated in the study and provided all the information needed to complete it, and we hope that this research will shed light on the great and pivotal role that these women play in the difficult circumstances facing them and as a model for other Syrian women. We hope these new roles will enable them to play a more positive role in society now or later when the emergency situation that has forced them to assume these difficult responsibilities and roles is no longer in place.

Part One –
The Economic and Social Conditions of Female Handicraft Workers in Damascus and Its Countryside

The change in work-related gender roles experienced by Syrian women is linked to a range of variables, some of which are directly related to the war and its immediate consequences, while others date from before the war. To clarify the gender variables the study focuses on, and to understand the past and current choices of the women we surveyed, it is necessary to describe their current and past living conditions and the relevance of those conditions to the women's practical choices, such as their educational and family background and their looking after themselves, children or other family members.

Several factors contribute to the measurement of change in gender roles: whether women had a previous job and the nature of that job are important indicators upon measuring the magnitude of change, since previous employment argues against the supposed gender impact of war in overcoming customs and traditions. The reasons for choosing the current work, whether personal or compulsory, may reflect the extent to which the caused change is imposed or chosen. The educational level of women also plays a role in understanding the opportunities for working women in Damascus and its countryside in a period of war economy and after. The presence of a husband as breadwinner also affects women's attitudes toward work, as displacement and the resulting significant economic need in inflationary conditions push women to seek additional financial resources.

In the following paragraphs, we will review in detail the conditions of the women participating in the study at the level of work, marital status and children, comparing their situation before and after the onset of war in 2011, in addition to reviewing the conditions of displacement they are currently experiencing.

1. The Work

Having a previous job and the reasons for its selection are an indication of the social status of women and their environment. The women in the sample cover a range of situations. At one end are women whose male guardians completely prevented them from working, as work affects the stereotype that men are the breadwinners and the family's source of income. These are women who used to consider themselves as having no need to work because of the economic sufficiency achieved by men. At the other end of the spectrum are women who were previously doing work similar to their current occupations inside the home (such as beading, making sweets, and other handicrafts), work they chose because they wanted to meet their economic needs without experiencing lifestyle changes.

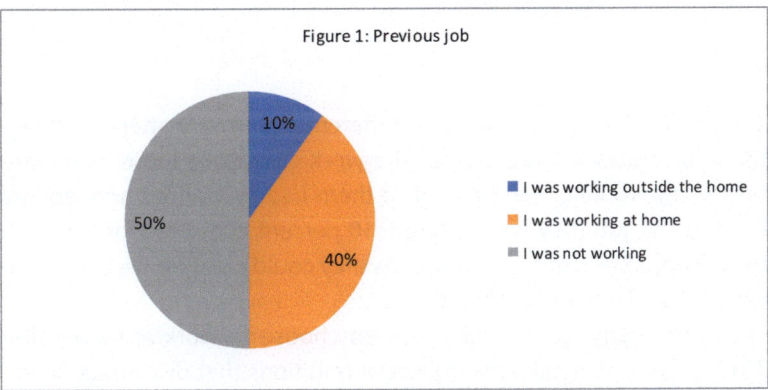

Figure 1: Previous job

Concerns about women spending long hours outside the house and the possibility of mingling with the opposite sex, along with other family members' inability to closely monitor women's behaviour, were among the biggest barriers that pushed women seeking to work to choose a job they could do in the home. The fact that a smaller proportion of them have previously worked outside the home is an indicator of an easing in the burden of strict social norms in their environments. Figure 1 shows the percentage of women in the study sample working outside and inside the home before 2011. Half of the sample did not work in the past, 40 percent used to work inside the home, and 10 percent worked outside the home.

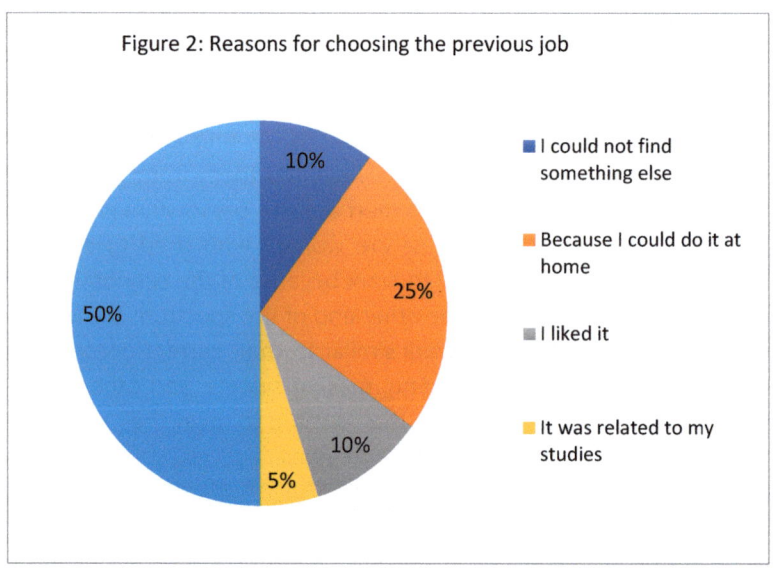

Figure 2: Reasons for choosing the previous job

The pre-2011 work options were different from those afterwards. Figure 2 illustrates the reasons for the work choices of those who were previously working. Ten percent of them had no options and worked in whatever job they were offered, 10 percent chose to work in a job they liked, 25 percent preferred jobs they could do at home, and 5 percent did a job related to their studies.

There are many reasons why women choose to work at home after 2011, such as the still powerful social traditions that discourage women from leaving their homes and mixing with others at work, in addition to the task of looking after children and the lack of caregivers if mothers leave their homes. Women sometimes choose home-based jobs for security reasons, such as the fear of moving to distant places under war conditions, fears of risks such as kidnapping, rape or even murder, and fears of facing interrogations at checkpoints and other inconveniences[5]. According to the International Committee of the Red Cross (ICRC), women's empowerment needs a different approach to the concept of women's vulnerability. In 2007, the Committee reviewed the assessment of vulnerability to improve its intervention

---

[5] For example, at checkpoints, women wearing headscarves who are from opposition areas are subjected to greater scrutiny compared to others.

methods in a report titled "Women in War: A Particularly Vulnerable Group?"[6]

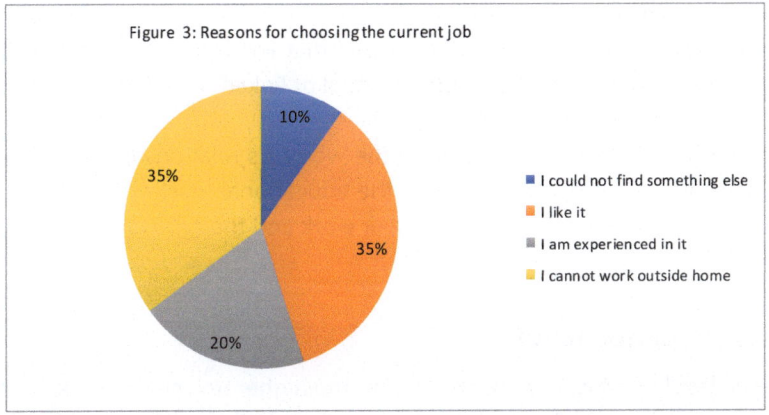

Figure 3: Reasons for choosing the current job

Figure 3 shows the reasons for choosing the current job. Thirty-five percent of the respondents said that the reason is their love of the nature of this work, 35 percent said because the work is done at home, 20 percent said it was because of prior experience, and 10 percent said it was because of the lack of other opportunities.

Some women expressed additional personal and objective reasons for choosing their current job, such as professional vocational study, a sense of visual beauty in this job, a habit of being surrounded by women working in such jobs, or a close relationship with another female worker of the same social and material status. Good social relationships are formed among female workers in the same company or work group, and they share many life concerns, and thus they receive psychological and social support through their work as well.

It is worth mentioning that handicrafts are of an artistic nature and enjoy social acceptance as a women's craft by the communities in Damascus and its countryside. It is easy to start earning money quickly with such a job, as it does not need scientific qualifications like more technical jobs. Working and mastering handicrafts jobs requires only

---

[6]   International Committee of the Red Cross (ICRC). "Women in War: A Particularly Vulnerable Group?" [electronic reference]. ICRC, available at https://bit.ly/2VFUiVA (Arabic) and http://bit.ly/2k8QoHq (English), published on January 3, 2007, accessed on April 10, 2019.

training and manual skills. This is a factor that influences the choices of women working in these jobs in general and the women included in this study. However, these indicators are still able to give an impression of the relationship of these women with their work, the reasons for choosing it and the circumstances that led them to engage in it, whether these conditions were personal or linked to the general situation experienced by the country.

After reviewing the nature of the women's relationship with their work, we will proceed to review the educational level of the women and its impact on the nature of their work and the new roles assigned to them.

2. Education level

The level of education, particularly at the higher university levels, is an essential factor in Syrian women's practical choices and is an indicator of their mobility and freedom of decision-making. It is inversely proportional to the percentage of women who chose home-based manual labour, as scientific specialization opens a wider door to their employment options. It is important to note that some of the conservative communities in Syria, to which the women in our study sample belong, still do not favour university study for women because universities are far from their place of residence and because of the possibility of women mixing with men there. Social customs like these negatively affect women's choices and career prospects.

The educational level of the women in the study sample ranges from primary and higher school levels to intermediate university education. However, the majority of them have reached the intermediate, preparatory and secondary school levels. Seventy percent of the study sample have a high school diploma or intermediate institute certificate in the field of female arts. They earned these certificates from schools affiliated with the Directorate of Technical Education in the Ministry of Education, where they learned knitting, sewing, housekeeping and other skills.

Those who have reached the university level are still studying while working to help cover their university tuition fees. This indicates that they consider their job a temporary one and that they will be looking for alternative options as soon as they have a better chance. This group consists of unmarried women, as the educational level of women in Damascus and its countryside is linked to the problem of early marriage in local communities. Female students drop out of school early after they marry as teenagers, and this simultaneously prevents them from doing work outside the home.

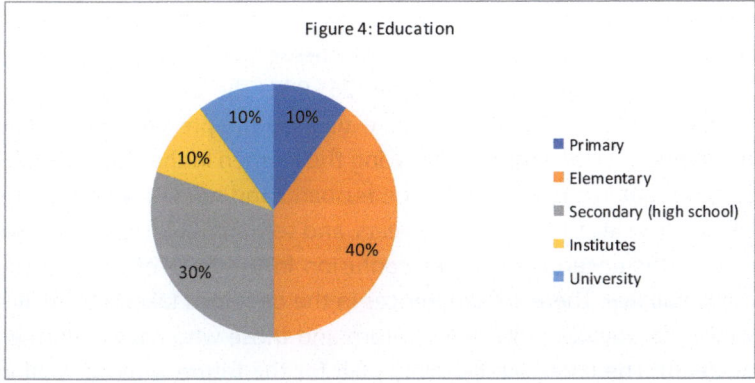

Figure 4 shows the educational levels of the women in the study. Forty percent of them reached the preparatory level, 30 percent reached the secondary level, 10 percent finished primary school, 10 percent obtained institutes' certificates, and 10 percent are studying at universities.

From these indicators, we note that most of the women participating in the study did not go beyond their secondary or preparatory education. This is a major factor affecting their choices in terms of work and life in general, especially in light of the current situation where most of them are in households that have lost their male breadwinner. The women thus became responsible for making the family's most important decisions, as we will see in the next paragraph discussing the women's marital status.

3. Marital status

Married women make up 90 percent of the study sample, however, in most cases the husband is not with the family. The loss of a breadwinning spouse is the main factor that has led most of the respondents to seek employment opportunities to support themselves and their families. The husbands are absent for a variety of reasons, including death from natural causes or as a result of hostilities in the areas of their origin.

According to statistics published in October 2015 by the Democratic Republic Studies Centre, a civil-society organization based in France, the number of missing Syrians then exceeded 109,535 persons, while the number of detainees exceeded 265,000. The Syrian Network for Human Rights, on the other hand, estimated the presence of 215,000 detainees in 2015[7]. From interviewing the women in the study sample, we note that in cases of loss of the husband and not knowing his fate, the family waits for months and years, and sometimes loses hope in his return. This uncertainty causes confusion in the form of life that the family will live. There are differences in the decisions taken by families waiting for their breadwinner's return and those who have confirmed his death. The latter can certainly plan for the future without waiting for him. This makes most of the solutions chosen by the female heads of household closer to the temporary; they rent a simple house, with little furniture, and do simple jobs without full or permanent commitment to their jobs.

There are no precise statistics on the number of missing men and detainees who have left their wives behind. However, a field study by the UN High Commissioner for Refugees (UNHCR), titled *Woman Alone: The Fight for Survival by Syria's Refugee Women*[8], spoke of more than

---

[7] See the following reports:
Syrian Network for Human Rights, *2015 Detainees in Syria* [electronic reference], available at https://bit.ly/2vCZIp2, published on January 5, 2016, accessed on April 10, 2019.
Democratic Republic Studies Center, *Comprehensive Statistics of Victims of the Syrian Regime Crimes Through the End of October 2015* [electronic reference], available at https://bit.ly/2WqCe2g, published on November 6, 2015, accessed on April 10, 2019.

[8] United Nations High Commissioner for Refugees (UNHCR), *Woman Alone: The Fight for Survival by Syria's Refugee Women* [electronic reference]. Available at

145,000 Syrian refugee households in which women were running their homes alone, and had to take control of their families' matters after the loss of men.

The existence of a male breadwinner is linked to the family's economic management and responsibility for the children. These responsibilities are automatically transferred to women in the family once the men are missing, putting women in a relatively new position and forcing them to address its economic consequences through manual labour.

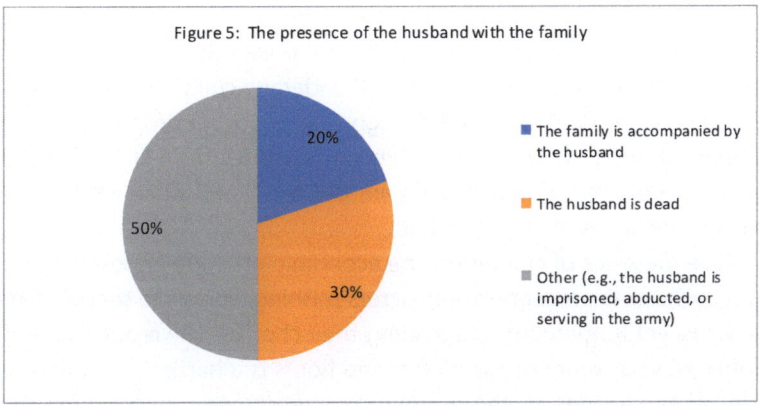

Figure 5 shows the distribution of families according to whether the husband is present. In 30 percent of the families studied, the husbands were dead. In 50 percent of the families, the husbands were absent for other reasons, such as imprisonment, abduction, or compulsory military service, while 20 percent of the families were accompanied by the husbands.

In addition to the absence of male breadwinners, the presence or absence of children is another key factor in women's recourse to employment, and factors like the number and age of the children impose specific limits on women's options. In the following paragraphs we will discuss the status of children for women who participated in the study.

---

https://www.unhcr.org/ar/53bb8d006.pdf, published on July 8, 2014, accessed on April 10, 2019, p. 7.

4. Children

Having children—or not—obviously plays a big role in the need to find an income source, as an increase in the number of family members will increase the basic needs for sustaining life and therefore the need for women to work. Adult offspring contribute to covering the family's expenses and sons may partially become breadwinners, while having to take care of minor children imposes on women a more urgent need to meet the childcare requirements.

The women participating in the study were the breadwinners for their children, if any. In households with children, the distribution was as follows: 80 percent were minors and looked after by the mother alone, and 20 percent were adults. All underage children were currently attending school, despite the high rate of school dropouts among displaced families. According to recent statistics from UNICEF, more than 2.1 million children in Syria are out of school, while another 1.3 million are at risk of dropping out[9].

The presence of children in the economic situation imposed by the war is one of the most pressing factors pushing women to seek income sources, yet fundamentally affecting their choices. The requirement of some jobs to commit to work for long hours is a barrier to women, as are other changes in the women's circumstances, such as moving away from the larger extended family that can contribute to childcare if they are in their immediate space, especially since most women are displaced from their home areas because of the war. In the next paragraph, we will review the conditions of displacement experienced by these women.

5. Displacement

The displacement of families has a pivotal impact on their economic situation. The families that have fled the war lack basic necessities of life, making it urgent that they find income sources and secure their needs. Thus, securing an income by the women becomes imperative, taking priority over certain traditions that prevent some males from

---

[9] United Nations Children's Fund (UNICEF), *Humanitarian Action for Children 2019 – Syrian Arab Republic* [electronic resource], available at https://uni.cf/2jYrl4e, posted on June 12, 2019, accessed on August 27, 2019.

accepting women's right to work for reasons of stereotypical presumption about the male's role in supporting their families.

This study was carried out in towns of the Damascus countryside (Jdaydet Artooz) and Damascus Governorate (Masaken Barzeh, Barzeh al-Balad and Rukn al-Din), which are located on the outskirts of Damascus and are often inhabited by low-income workers, where the women in the study sample live. The women were divided into those who originally lived in these places, and others who were displaced to these places after 2012 from the Eastern and Western Ghouta region in the Damascus countryside.

They all share the same difficult economic and social conditions, often with the problem of the husband's absence, and hence the absence of a male breadwinner. However, we noted the presence of assistance from the surrounding community (at least moral support) when the woman stays in her home and familiar social surroundings, while displaced women suffer from moving away from the rest of their family for economic or security conditions.

Displaced women constitute 80 percent of the study sample and are those with the greatest need for a source of income secured through their handicrafts work and other jobs done irregularly. Twenty percent of the women surveyed are still in their primary place of residence before 2011.

Looking at these factors we have reviewed regarding the women's work, education level, marital status, children and displacement, we note that the economic and social conditions have pushed women to work, while the absence of the husband turned the head of the family hierarchy to the mother. Displacement facilitated changes in the women's dress and necessitated other changes, like mobility and dependence on themselves in the conduct of personal and family affairs. Thus, resorting to work, because of the above-mentioned living conditions, is a form of decision-making, which was not a matter for women in the past, but a decision taken by the men in the family. Thus, women's practice of handicrafts has led to gender-role changes. In the next section of this report, we will explore the parameters and the mechanism of these changes and their impact on the stereotyped image of women in conservative Syrian communities.

Part Two – Monitoring the Gender Variables

We can see the change in the society's stereotypical ways of thinking about the role of women after 2011. Before, men played the role of breadwinner and women depended on them. After the onset of the war, the lack of this male breadwinner and the fact that women have had to rely on themselves demonstrated a flaw in this stereotyping, especially since women were not previously considered qualified for work and self-reliance. In addition, the motivation behind women's working now varies from the possibility of assisting the husband in the task of securing the household expenses to performing this task alone.

Women confronted this radical change in ways of thinking and became breadwinners for themselves and their families, gaining a rapid development in the sense of ability to act in general. This formed a strong basis for taking charge of other personal decisions that were interfered with by other family members before.

In this section, we will try to explore the details of the changes in the gender roles of the women participating in the study through three factors: their freedom of movement, housing conditions and the responsibility for taking care of children. We consider these factors to be subjects of direct significance in assessing these changes. In order to monitor a change in stereotypical roles, the types of these roles must be determined first. Perhaps the right to self-decision for women is the umbrella from which the freedoms of dress, movement, choice of housing, employment and other personal choices all follow, and these freedoms are the essence of the stereotypical difference between males and females in the society to which the women belong.

1. Freedom of Movement

Freedom to leave the home is an essential factor in monitoring gender-role change in Syrian society. For women in the social circles represented in the study, strict traditions prevent women from going out on their own unless it is with the consent of the father, husband or another male guardian, and often a male relative's company is required. Justifications for going out of the house are required, and the purchase of personal needs, specifically with regard to dress, is often the most compelling justification for going out, except in some cases in which the exit of women from the house is generally forbidden, especially in the more conservative circles, which find justification for this attitude in some religious teachings[10].

The possibility of travelling alone is out of the question because of reasons related to male guardians' fear for the safety of women and fear of the women's exercising emotional or sexual options that may not appeal to them. This prohibition sometimes takes the form of religious commitment. Islamic Shari'a is used as a final deterrent to women's debate about their right to move freely. This use is based on strict jurisprudential texts that are limited to the studied society in Damascus and its countryside[11]. While women are allowed to drive, it is almost impossible for them to do so when the family owns one car only and the family includes a man who drives. If so, this man is often entrusted with the task of picking the family women up.

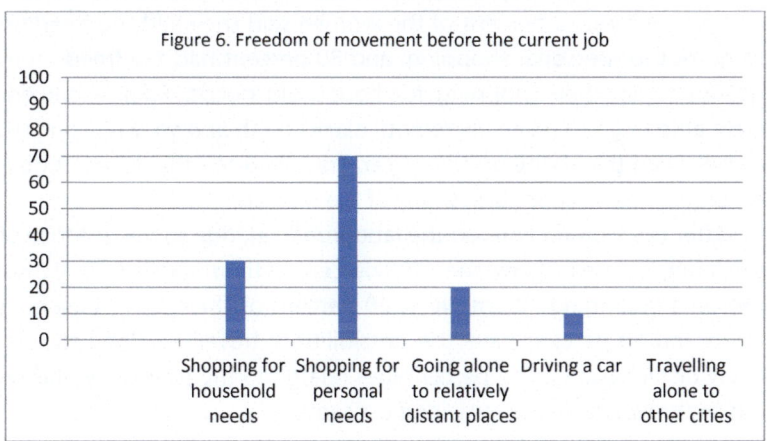

Figure 6. Freedom of movement before the current job

[10] A famous Levantine proverb says: "A woman, from her family's house to her husband's house and from there to the grave." That is, it is preferable for a woman not to leave the house of her family except to go to her husband's house and stay there until she dies.

[11] Abu Sa'id al-Khudri reported that the Prophet Mohammed said: "A woman should not set out on two (days') journey, but in the company of a Mahram." See Ahmad al-Zayn Hamza, *Musnad Ahmad bin Hanbal,* Volume 10, Cairo: Dar al-Hadith, 1994, p. 110.

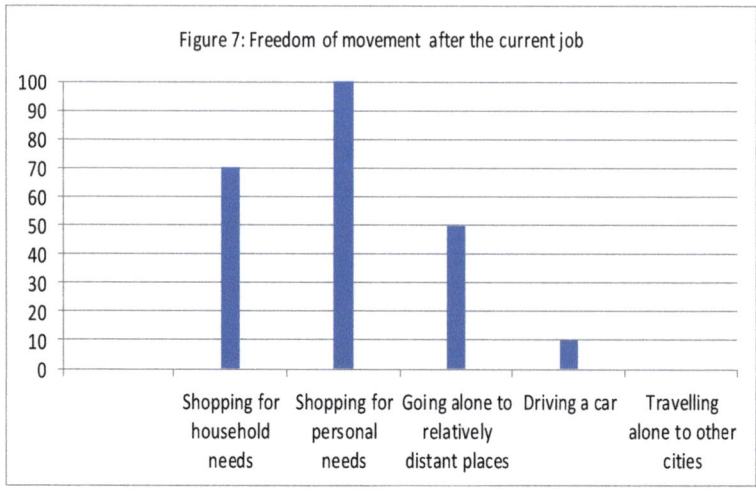

Figure 7: Freedom of movement after the current job

Figures 6 and 7 show the differences in the freedom of movement for the women in the study sample before and after they took their current job. Before, 70 percent of the women said they had the freedom to go out for personal shopping, and 30 percent had the freedom to go out to do their shopping for household needs. Twenty percent were able to go to relatively remote places on their own, while only 10 percent had the ability to drive a car. Travelling to other cities or governorates was not possible for any of the women.

After the women entered the labour market, due to the absence of the family's male breadwinner, 100 percent of them said they buy their personal belongings themselves, 70 percent of them buy household items and 50 percent have the possibility of travelling alone to relatively distant places. The percentages related to the possibility of driving and travelling alone remained the same.

For the women who participated in the study, women's mobility alone plays a logistical and social role in ensuring their ability to meet the family's needs. We note that all women currently do personal shopping and do the shopping for their children. However, 30 percent of them do not currently shop for the household items, for social reasons related to the idea that this task must be entrusted to a male in the family. That male may be the father, son or brother, even if this son or brother is a minor. This is not affected by the fact that they pay for these supplies from the woman's own money.

For the women in the families studied, having a car became more difficult with the deterioration of the economic situation, not to mention the difficulty of learning to drive in the context of wartime economic and social situations. Meanwhile, the possibility of travelling alone remains the same—that is, impossible, as travel becomes more difficult for security reasons, the presence of military checkpoints and the fear of kidnapping.

2. Housing

Before the repeated displacements experienced by some families, it was not common for a woman to live with her children in a separate home alone in the absence of the husband, in case of his death, and particularly if the children were minors. In fact, the woman often moved to live in her family's or her husband's family's house, in cases when she and her husband were living independently in a separate house before his absence.

Now, the displacement of entire families and the social and religious conditions that make it more difficult to accept the presence of a number of people in one house, as well as the full responsibility of the women to take care of their children, have played an increasingly important role in leading the society to accept the women's independent lifestyle step by step.

A report by the International Committee of the Red Cross titled "Women in War" states that women "show remarkable strength, as can be seen from their roles as combatants or peace activists and the duties and responsibilities they take on in wartime to protect and support their families. […] They often find ingenious ways of coping with the difficulties they face when fulfilling the role of head of household, caring for and earning income for their families or taking part in community life"[12].

Work contributed to securing the needs of the women who participated in the study and alleviating the new economic burden imposed on them without relying on the family, even if only in part. Independent housing increases their financial responsibilities, while sharing a house with the rest of the family reduces the financial burden of high

---

[12] ICRC, "Women in War: A Particularly Vulnerable Group?"

rent. Nevertheless, half of the women in the sample preferred living independently over living with the larger family—a step that is likely to move towards more independent lifestyle choices—and were willing to pay for it.

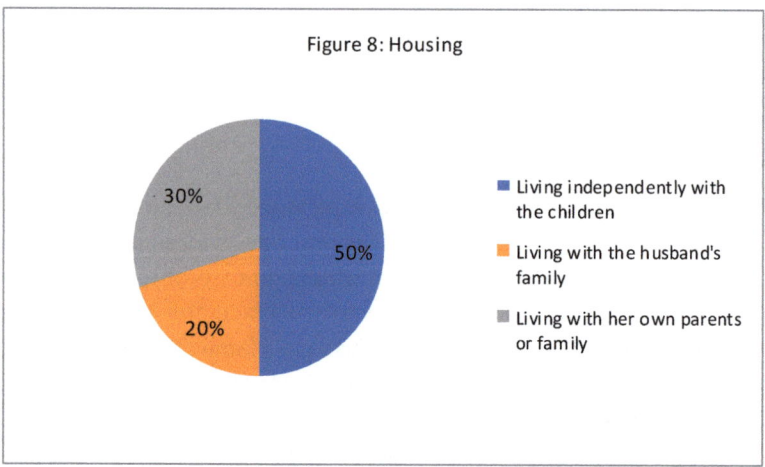

Figure 8 shows the current housing conditions of the women in the study sample, where 50 percent of the women live in a separate residence with their children, 30 percent of them live with their children and the mother's parents, and 20 percent live with their children and the husband's parents, even if the husband is absent. This is a development compared to the pre-2011 situation, when most of the women living with their children would have found it difficult to live in a separate dwelling if the husband was lost or deceased or if the couple were divorced. Today, displacement, the poor conditions and small sizes of rented houses, and the large number of individuals living in them, in addition to the many cases in which the wife stays with her children without the husband, facilitate getting the parents' approval for a woman's decision to live in a separate dwelling. This gave women the opportunity to obtain greater independence in making decisions at the level of home management, child rearing and other responsibilities, which are often considered primarily the husband's responsibility. The most prominent of these responsibilities are the way of bringing up the children and taking full responsibility for them, which is often fully transferred to the mother. This will be discussed in the next section.

3. Responsibility for Raising Children

According to customs, traditions and the Syrian law, the mother has the right to custody of her children up to the age of 13 years for boys and 15 years for girls. She will also receive maintenance from the father or his parents in case of his absence. The custody period is also a period of supervision of the child's daily life details. Article 137 of the Personal Status Law requires a person to be an adult and have the mental ability to maintain the child's health and morality as prerequisites to be eligible for the custody[13].

Such legislation and the fact that women stay at home with their children for a long time and their responsibilities regarding the children's education have not fundamentally affected the economic responsibility. The father, if present, has the greatest role in fulfilling the needs of the children and imposing his economic decisions on them as the breadwinner of the family. There is a clear change, however, in the way families deal with a missing or deceased parent regarding the guardianship of children. Laws stated that custody will be taken from the mother in case of her marriage, for example, while in the current situation, the mother's parents or her husband's parents generally prefer for the children to stay with their mothers, due to the difficult economic situation and the inability to take care of a large number of children. This is especially true as most families are poor or middle-class and do not have the financial ability to foster grandchildren. Moreover, a large proportion of families have more than one missing son and more than one family to support. This makes it better to have each group of children with their mother and to ignore adherence to customs as before.

---

[13] Syrian Arab Republic, People's Assembly, "Law 59 of 1953, Personal Status Law" [electronic reference], on the Syrian People's Assembly website, available at https://bit.ly/2ltW20o, accessed on April 10, 2019.

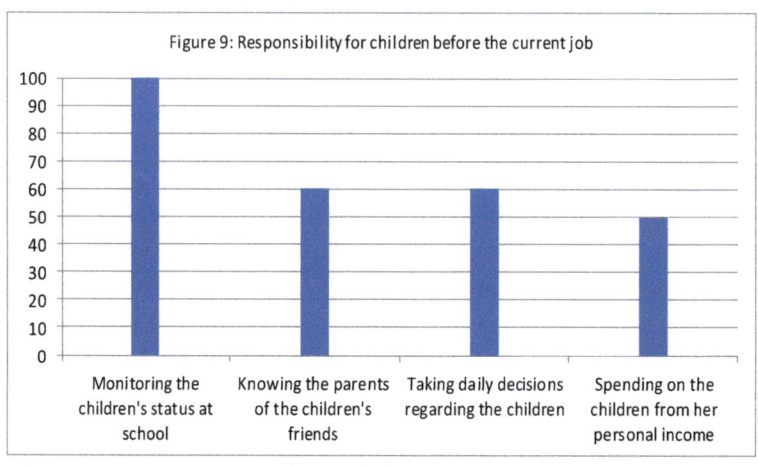

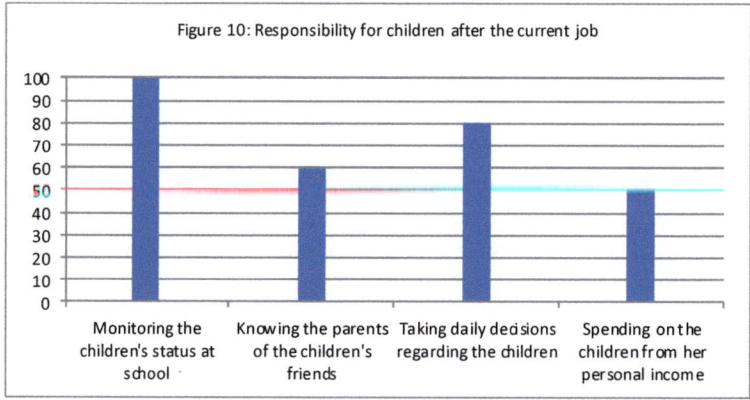

Figures 9 and 10 show the changes in the responsibility of the women included in the study for their children before and after taking their current job. Before, 100 percent of the women were responsible for monitoring the status of the children in school, 60 percent of them personally knew the parents of their children's friends, 60 percent of them used to make daily decisions regarding their children, and 50 percent of them spent on their children from their own income.

After taking the current job, it is still the case that 100 percent of the women are responsible for monitoring the status of children in school, 60 percent of them personally know the parents of their children's friends, and 50 percent of them spend on children from their own income.

In one significant change, however, we note that a larger share of the women, 80 percent, now make daily decisions regarding their children. This is an area where the father was more involved before.

We also note that the proximity of the larger extended family opens the door for intervention by paternal and maternal uncles or grandfathers in decision-making regarding the children. In in-depth interviews, women expressed the difficulty of avoiding the interference of relatives in their children's personal lives, particularly after the father's absence. Social custom encourages other family males to intervene in every detail of the lives of children, particularly in the case of economic responsibility for them or at least in covering their expenses. This is true even if the sons live with their mother in an independent house.

Today, all the women who participated in the study support their children from their personal income, whether they were living with them in a separate home or in the family home with their parents. Income from their jobs may not be enough to cover the children's expenses, and women may turn to other sources of money, such as parents, aid, charities, and savings.

The fact that women are in charge of economic responsibility has given them more space to make decisions regarding their children, specifically in terms of making decisions that require spending from their own money (on enrolment in a particular course, hobby development, expense priorities, clothing, personal needs like haircuts, summer clubs, and so forth). By 2015, the percentage of Syrian households headed by women was estimated at 12 to 17 percent, up from 4.4 percent in 2009. In 2018, that percentage had risen to 22.4 percent.[14]

As for the current economic and social conditions, we have noticed that the practice of customs and traditions is gradually changing to cope with the Syrian family's current situation. In addition, women have access to quick legal solutions from judges to facilitate the

---

14 "Barriers Dropped by Women and War: A Paper on Syrian Women and Their Economic Role" [electronic reference], *Al-Hal* website, available at https://bit.ly/2MIFPDI, published on January 23, 2018, accessed on April 10, 2019.

movement of mothers and their children, and transfer custody to them in case of loss of communication with the father for some reason. Some 70,000 temporary legal guardianships were documented in Syria in 2017[15]. These legal solutions remain temporary but they serve to make women's lives and movement with their children more manageable under the current situation.

The women in the target group for this study come from rural areas where there was a low level of higher education because of social customs that hindered women's opportunities to study. In May 2017, Bareeq Society for Education and Human Development conducted a survey of Syrian women over the age of 18 inside and outside the country. It found that 81 percent of the 1,006 women surveyed said that social norms in Syria impeded women's success[16].

Besides the social reasons hindering educational attainment, there are also economic ones related to the preference for children to work in agriculture and other sectors (a good percentage of the women in this research own an agricultural plot of land), in addition to the presence of many industries in Damascus countryside and the spread of factories into the region. The countryside of Damascus is teeming with factories and installations for all industries, with about 19,000 factories and plant facilities of various types, according to Wikipedia[17].

Through in-depth interviews, it was found that changing circumstances and the sudden loss of homes, factories and employment opportunities made women more persistent and more determined than ever before to educate their children as a way to secure a good future for them. Education provides permanent and renewed employment opportunities, in their opinion, and makes their children better able to cope with difficult circumstances (such as those experienced by the mothers now). Education decisions were previously determined primarily by the father's wishes, while the mother's wishes for the future

---

[15] "One Thousand Temporary Legal Guardianships This Year, Most of Which Are Travel Permits for Wives Claiming Their Husbands Are Missing in Order to Travel" [electronic reference], *Al-Watan* newspaper, available at http://alwatan.sy/archives/128044, published on November 20, 2017, accessed on April 10, 2019.

[16] "Barriers Dropped by Women and War: A Paper on Syrian Women and Their Economic Role".

[17] Wikipedia (Arabic), "Damascus Countryside" [electronic reference], available at https://bit.ly/1EeOaJo, accessed on April 10, 2019.

of the children were secondary. Today, this has turned into a decision taken by the mothers and a planning for their children's lives and personal choices. These aspirations about the future of children predict the women's views of society as a whole and their implicit desires towards themselves as well.

All of the women in the study sample said that before and after their current job, they did not differentiate by gender in their preferences for their children's future education. This contradicts the stereotypical picture of mothers' wishing to get their daughters married at an early age before they reach higher education. Although the pre-2011 answers about preferences for the future of the children cannot be verified, the responses still mean that the circumstances of the war and the changes the women have experienced as a result of the war have contributed to excluding the traditional gender-differentiated way of thinking regarding the necessity of education, which is often considered a primary need for males and a secondary one for females, who are prepared instead for early marriage and becoming housewives.

The change in the factors we have mentioned (movement, housing, and responsibility for raising children) has led to a change in the lifestyle of the women who participated in the study. This, of course, has led to a change in their personalities, ideas, and awareness that they can play a greater role in determining their own lives and their children's, even if the emergency situation in which they now live fades away. Did this change in gender roles really happen? Is it permanent or will it go away once the current situation is over? These are questions we will try to answer in the following paragraphs.

4. Desire for Change

During the interviews, the women expressed a clear change in their lifestyles and ideas about their rights and duties. Whereas previously they were women who did not go out of the house and felt surprised by a social atmosphere in which women worked outside the home and provided their household needs, they are now women who take on major responsibilities, sometimes for an entire family. The interviews showed satisfaction and pride among the women who participated in the study for having at least enough private income to meet their personal expenses. Most of the women expressed a desire to continue working whatever the circumstances of life, even if their financial conditions improve.

Eighty percent of the women believe that work is not a primary cause of their ability to live separately from their parents or their husbands' families. Rather, they see their independent housing as due to factors like the living conditions of the parents, displacement and forced relocation in general. Jobs, however, help make such a change possible. Thirty percent of the women started working while living in shelters and jobs later helped them find a place to live outside these centres.

All of the women previously linked work to their material needs. After working for a long time, they started to link work with social status and independence as well. Thirty percent of them expressed unachievable wishes that they would be financially sufficient to stop working, due to lack of time, fatigue from taking on responsibilities alone, and difficult housing conditions such as poor lighting, narrow spaces and having many responsibilities due to their living with parents. However, they expressed their desire to continue in handicrafts as a hobby to fill the gap when their financial and living conditions improve. The rest of the women linked work to positively organising time.

It is also worth noting that the household chores do not go away when women take on new workloads, causing some women to complain about and resent the changes in their lives. For example, in her 1949 work *The Second Sex*, Simone de Beauvoir cited a study of female workers at a Renault factory in France that suggested that many of the women would prefer to stay at home rather than work at the factory. "There is no doubt that they have achieved economic independence only within an economically impoverished class," de Beauvoir wrote. "And besides, the tasks they perform in the factories do not relieve them from the burden of household chores. If they had the chance to choose between spending forty hours at the factory or at home, their answers would have been different"[18].

Ninety percent of the women in our research agreed that they would not want to spend a long time without work, regardless of the

---

[18] Simone de Beauvoir, *The Second Sex*, Arabic translation by Sahar Said, Damascus: al-Rahba Publishing and Distribution, 2015, p. 234.

financial return, because of boredom, or a sense of emptiness and futility. They reported that, before they worked for money, they spent a lot of time doing simple manual work for the home or family to fill in the emptiness they experienced. All of the women said that handicrafts improved their psychological state and helped them to have fun at home, especially in case of being unable to leave the house. They see these works as a way of expressing their feelings and personality and highlighting their specific skills and talents.

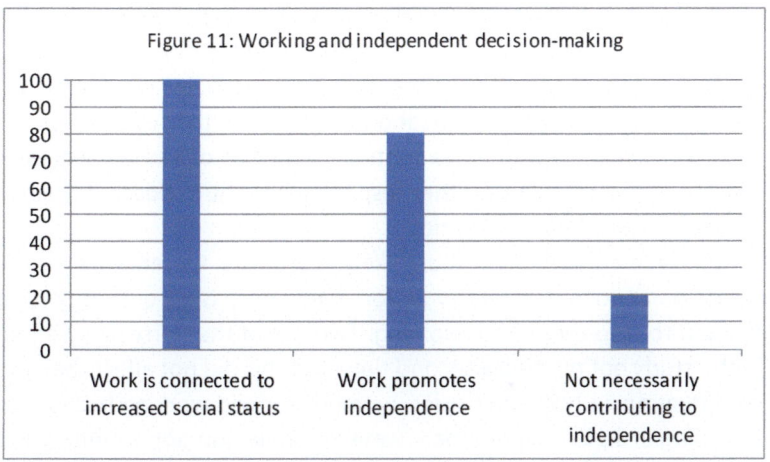

Figure 11 reflects the relationship between work and decision independence. All of the women associated work with increased social status, and 80 percent said work gave them more independence. Twenty percent of the women said they believed that work helped in increasing their independence of decision-making but was not a direct reason for it, because the job carried out by these women is performed at home, and social independence is mainly linked to customs and traditions and not only material independence.

After reviewing the changes experienced by the women at various levels, we re-ask the question: Are they agents of change or objects affected by it? These changes have actually transformed women into agents of change in their societies, but this has not been caused by their own propensity for seeking change, but rather by external factors linked to the surrounding circumstances that have driven them into it. This raises the question of whether these changes are desirable and will persist, or whether they are enforced by necessity and whether it is possible that women will return to their previous status when the reasons for change no longer exist. Will women return to their former lives in terms of making decisions for themselves and their children? Will their previous ideas on their rights and duties be restored if one of the variables, such as displacement or the presence of the male breadwinner, goes back to its previous status?

Recognising the value of work and linking it to social status is the most fundamental factor in this change in attitudes about gender, and is the clearest consequence of all other variables that have brought Syrian women into a space where they are effectively taking responsibility. Perhaps it is the continuation of this work that establishes the special criterion for women to continue to obtain or renounce greater privileges and to give up other responsibilities. It is closely linked to women's equal rights and duties.

Today, marriage is the most likely reason for the absence of the need for a woman to work, or for males exercising authority over women to stop them from working. Two out of the thirty-five women in the study got married a second time, but this did not affect their job continuation. In two other cases, women stopped working because of resettling in a place where there were no similar job opportunities and because of a return to a family that covered the woman's financial needs.

In any case, the overall economic situation of the respondents remains poor to an unseen extent, leaving most of them in need of work and finding an urgent financial resource. This keeps open the question about women giving up the right to work if their previous life conditions return, or if it is possible to reverse the change in their lifestyle at the level of independence and decision-making, and to reverse their gains in this regard.

Conclusion

This paper sought to monitor gender-role changes in a sample of Syrian women whose circumstances changed after 2011 due to the ongoing war in Syria, and in Damascus and its countryside in particular. These women entered the labour market in a timid manner that did not completely change their lifestyle yet greatly affected it. Additional factors resulting from the war and the accompanying economic collapse have contributed to this change. Displacement and the loss of a husband or other male breadwinner were among the most important causes for women in the study to enter the labour market, besides a set of secondary variables related to their environment, educational and professional backgrounds.

To monitor these changes, we started from a key question about how these changes really affected the lives of women in the labour market. To answer this question, we relied on resources such as face-to-face observations, in-depth group interviews and questionnaires, working with a sample of 35 female handicrafts workers between the ages of 25 and 55 who work in their homes.

Economic and social conditions have led the women to work, while the absence of a husband has shifted the head of the hierarchy in the family to the mother. Displacement has facilitated changes in dress and necessitated other changes in mobility and self-reliance in the conduct of personal and family affairs. Thus, choosing to work due to the living conditions was a form of decision-making, which was not a matter for women in the past but something decided by the males in the family. Women's practice of home-based manual labour has led to gender-role changes that can be seen in many aspects of their lives, compared to how they lived before they assumed these new responsibilities and roles.

Women responded to this radical change in ways of thinking by becoming breadwinners for themselves and their families, acquiring a rapid development in the sense of being able to act in general, and forming a strong basis for taking control of personal decisions that previously were interfered with by other family members. Women have gone beyond the problem of being able to socialise with other segments of society in the context of what their work entails, and outside the home in general, and now have a greater sense of being able to act, interact with and participate in the construction of the world proudly. The change in women's stereotypical roles has not led to a direct or significant change in the customs and values that govern gender relations; however, it has affected women's authority and ability to make decisions about themselves and their children.

Gender-role changes in self-esteem remain uneven among women, and it's not clear whether these changes will endure, which requires follow-up and further study over the coming years. The poor economic situation, urgent need for jobs and this change in the relationship between Syrian men and women on the basis of interests are increasing in priority day by day as a long-term consequence of the war in Syria, and this should not be overlooked by later studies.

Although it is difficult to select and identify the variables governing the change in the relationship of the women who participated in the study with themselves and with the world over the past years, especially since the factors leading up to these changes have not yet disappeared, trying to understand these changes is necessary to overcome the negative stereotypes of males and females in Syrian society, and spread the need to enjoy equal rights, including duties through which they all contribute to serve the community's development.

Perhaps these changes and a growing interest in politics and public affairs at the grassroots level are among the few benefits of all the losses that have befallen Syrian men and women. Studying and considering them as gains might be a duty for those who are interested in advancing Syria towards practising effective citizenship.

Finally, we would like to thank the women who inspired us regarding the importance of this paper, and generously agreed to participate in conducting the interviews and taking the questionnaires. I would also thank my friend the scholar Malak Shanawani for her support and observations to accomplish to this paper. Of course, the greatest thanks are due to the supervisor of this research, Dr. Hassan Abbas.

References

Books

Hamza, Ahmad al-Zayn. *Musnad Ahmad bin Hanbal,* Volume 10. Cairo: Dar al-Hadith, 1994.

De Beauvoir, Simone. *The Second Sex.* Arabic translation by Sahar Said. Damascus: al-Rahba Publishing and Distribution, 2015.

Bourdieu, Pierre. *The Weight of the World: Social Suffering in Contemporary Society.* (Arabic translation). Damascus: Kanaan Publishing House, 2001.

Hamadeh, Najla. Lentils and Caviar: Biographies of Women Who Live in Lebanon: Self-Construct, Agency, and Independence. Beirut: Arabic Scientific Publishers, 2015.

Online Resources

"Barriers Dropped by Women and War: A Paper on Syrian Women and Their Economic Role" [electronic reference]. *Al-Hal* news website, available at https://bit.ly/2MIFPDI, published on January 23, 2018, accessed on April 10, 2019.

"One Thousand Temporary Legal Guardianships This Year, Most of Which Are Travel Permits for Wives Claiming Their Husbands Are Missing in Order to Travel" [electronic reference]. *Al-Watan* newspaper, available at http://alwatan.sy/archives/128044, published on November 20, 2017, accessed on April 10, 2019.

United Nations Children's Fund (UNICEF). *Humanitarian Action for Children 2019 – Syrian Arab Republic* [electronic reference]. Available at https://uni.cf/2jYrl4e, posted on June12, 2019, accessed on August 27, 2019.

National WWI Museum and Memorial. "Women in World War I" [electronic reference]. Available at https://bit.ly/2Kyis3h, (n.d.), accessed on February 10, 2018.

Official Publications

Syrian Arab Republic, People's Assembly. "Law 59 of 1953, Personal Status Law" [electronic reference], on the Syrian People's Assembly website, available at https://bit.ly/2ItW20o, accessed on April 10, 2019.

Syrian Network for Human Rights. *2015 Detainees in Syria* [electronic reference], available at https://bit.ly/2vCZIp2, published on January 5, 2016, accessed on April 10, 2019.

Democratic Republic Studies Centre. *Comprehensive Statistics of Victims of the Syrian Regime Crimes Through the End of October 2015* [electronic reference], available at https://bit.ly/2WqCe2g, published on November 6, 2015, accessed on April 10, 2019.

International Committee of the Red Cross (ICRC). "Women in War: A Particularly Vulnerable Group?" [electronic reference]. ICRC, available at https://bit.ly/2VFUiVA (Arabic) and http://bit.ly/2k8QoHq (English), published on January 3, 2007, accessed on April 10, 2019.

UN High Commissioner for Refugees (UNHCR). *Woman Alone: The Fight for Survival by Syria's Refugee Women* [electronic reference]. Available at https://www.unhcr.org/ar/53bb8d006.pdf, published on July 8, 2014, accessed on April 10, 2019.

Appendix

The questionnaire used in this research and the questions asked during the interviews conducted with members of the study sample

- Name:
- Age:
- Place of residence:
- Previous place of residence (if any):
- Marital status:
- The husband:
- Children, their age and sex:
- People living in the same residence:
- Previous work:
- Current work:
- Date of starting work:
- Type of your job:
- Reason for choosing this job:
- Were you working before? And why?

Before you started working:

- Were you going out to do the shopping?
- Were you shopping for your personal or family needs at relatively distant locations?

- Did you have any experience in driving a car?

- Were you travelling alone?

- Were you going to your children's school to check on their progress in their studies? Or if there was a problem?

- Did you know the parents of your children's friends?

- Were you making decisions regarding your children? Who was sharing this task with you?

- What would you prefer for your children, work or study?

- What would you prefer for your daughters, marriage or study?

- Were you allowing other people to interfere in matters regarding yourself or your children (apart from your husband)?

- Who were these people?

- Did you think that a man was solely responsible for securing the family's income?

After becoming a working woman:

- Do you go out shopping?

- Do you go shopping your personal or family's needs at relatively distant locations?

- Do you have any experience in driving a car?

- Do you travel alone?

- Do you go to your children's school to check on their progress in their studies? Or if there is a problem?

- Do you know the parents of your children's friends?

- Do you make decisions regarding your children? Who shares that responsibility with you?

- What do you prefer for your children, work or study?

- What do you prefer for your daughters, marriage or study?

- Do other people interfere in matters regarding yourself or your children (apart from your husband)?

- Who are these people?

- (If yes) Do you still think that the man is solely responsible for providing the family's income?

In-depth interviews focused on the details of the ongoing changes and the form of daily life under such changes. These interviews allowed the respondents to express satisfaction or resentment. They also expanded on answering the questionnaire questions.

Small-group interviews included extensive discussions on livelihoods and how the women's relationships with themselves, their children and the surrounding community changed, before and after 2011, and before and after their going to work.

Ola Alshikh Hassan

Born in 1983, Alshikh Hassan studied French Literature at Damascus University. Since 2015, she has been working as a Production Manager at Sama Handicrafts, and since 2013 in Threads of Hope-Aghbany. She has also worked for various art and crafts companies.

The Damascene Textile Industry in the Cauldron of War

By: Zeina Shahla

Under the supervision of Hala Charif Alchach, PhD

Acknowledgements

Over the course of many months, I have sunk into the alleys, quarters, khans and old markets of Damascus, trying to explore the mysteries of this city's textile crafts—something I do not claim to have known much about before. Many people have opened the doors of their shops and workshops to me. I spent long hours listening to the sound of hand looms weaving colourful rugs, or of wooden templates stamping patterns on fabrics that will be embroidered later on.

I would like to thank all those who helped me to obtain any information about this research, no matter how simple it was, for it was very helpful to me to understand the entire scene, to increase my knowledge and even deepen my fascination with Damascus and its treasures, and to think about the role we Syrians have to play to protect our country from the effects of war that may not be visible today but are likely to emerge later on. I would like to thank—or more correctly pay tribute to—the late Mr. Ibrahim al-Ayoubi, one of the professionals of the brocade industry in Damascus, whom I met at a small workshop in the Sarouja market in March 2018. He spoke to me with warmth, passion and sorrow regarding a profession he inherited from his father, a profession that, like other Syrian handicrafts, faced difficulties and challenges during the war and the years before. In November 2018, Mr. Ayoubi passed away, and I regret not being able to give him a copy of this research.

Great thanks go to my supervisor, Dr. Hala Charif Alchach, for I would not have completed this research without her follow-up, encouragement and belief in me and the importance of what we are doing, which is part of the efforts to preserve our Syrian identity and cultural heritage from the fires of war throughout our country.

Summary

This research provides an in-depth study of the status of Syrian traditional textile handicrafts in 2018, about seven years after the beginning of the civil uprising of 2011 and the outbreak of war in the country, through a detailed study of three examples of Damascene textile handicrafts: brocade, the Aghabani, and handmade rugs. In addition, the research examines the changes that have occurred at the levels of the labour force, raw materials, machinery, production lines and outlet markets, thus highlighting the most important difficulties and challenges experienced by the workers in these fields, who are doing their best to continue their work and protect their crafts from the deterioration that their industry, like other economic sectors, has suffered during the war and its aftermath.

Seven years of civil war across the entire territory of Syria has claimed the lives of hundreds of thousands of Syrians and destroyed most of the country's economic sectors, besides creating a situation of unprecedented human havoc caused by emigration and internal displacement, especially for young people. Despite the priority of work to restore the most vital sectors of the economy, such as healthcare, education and housing, it is also important to give attention to the protection of Syria's cultural heritage—both the material and non-material, or "intangible," cultural heritage—during the conflict because this heritage is connected to the country's identity and is one of its most important components. The destruction that has affected various aspects of this heritage during the war may take decades to rebuild.

This research is based specifically on the question of how the war affected traditional Syrian handicrafts as a component of the nation's cultural heritage. It also set out with the aim of trying to reach a set of conclusions and recommendations that may help, if adopted by experts and people interested in this field, to save these crafts from further loss or destruction.

We chose to focus on three Damascene textile handicrafts as examples for the study, namely brocade, the Aghabani fabric, and handmade rugs. The reasons for this selection are that these crafts are closely related to the Damascene identity and that sources capable of providing the relevant information are readily available. We studied the condition of these crafts before 2011, the year that formed a turning point in Syria, and the changes that took place in the chosen crafts during the war through 2018, the year of carrying out the study. This helped us make descriptive comparisons as much as possible between the status of these crafts before and after the war.

The research employed in-depth, face-to-face interviews with practitioners of these crafts at all levels, fields, and stages of production, as well as with workers in various workshops engaged in the sale of Damascene textile products, and related government officials. This entailed about twenty-five interviews. The research also benefited from a review of books, research and articles that speak either of the history of these crafts or of their current status. The research thus formed a combination of theoretical and field work.

The most important finding is that the biggest challenge facing the workers in the field of traditional Syrian craftsmanship is the lack of a cadre of qualified people who wish to learn these crafts and specialise in them, who are needed to replace the many professionals lost from these crafts as a result of emigration, displacement and death. There is a reluctance among the new generation to learn these crafts as a result of their difficulty and the low material returns they bring, compared to other crafts.

The second biggest challenge is the difficulty of selling the products of these crafts, both internally and externally. Many Syrians look on these goods as luxuries and are reluctant to buy them, given the deterioration of their financial conditions. There has also been an unprecedented drop in the number of tourists visiting Damascus, further limiting the number of potential customers. It is also impossible to export the products of these crafts abroad due to the lack of marketing strategies needed for that.

Other challenges include the destruction of many workshops, the loss of production machines that were damaged, stolen, or taken out of service for other reasons, and the difficulty and cost of securing raw materials. These are secondary challenges, according to the workers in the field, and can be overcome when the larger problems of labour, marketing and selling are solved.

Finally, the research produced a series of recommendations that were collected during the interviews, which we hope will contribute, in addition to the local efforts exerted in this field, to preserving the Syrian crafts from the danger of disappearing and being lost. There is an urgent need to apply these efforts and to place them within the priorities of cultural action in Syria. These recommendations include drawing up special plans to support the traditional crafts separately from plans to support other Syrian industries, giving attention to the workers in this field at the various professional, administrative and social levels, and in particular giving attention to the education of women in order to overcome the problem of Syrian society's loss of large numbers of men.

The recommendations also include linking these crafts with public and private institutions in other sectors, such as school education and higher education; working to spread public awareness of various aspects of Syrian heritage and the importance of protecting it and learning it; following effective marketing strategies to support craftspeople; and documenting everything related to these crafts.

Introduction

It is difficult to separate national identity from the various types of cultural heritage, both the material, such as the products of traditional crafts, and the non-material, such as the knowledge and skills used to produce those crafts. Cultural heritage is considered an essential component of identity and one of the most important factors that distinguishes one identity from others. It is also a solid foundation that immunises someone against the variables of life. It is like the pillars of identity, no matter how long the timespan it encompasses.

Heritage acquires an increased importance during periods of conflict that pose a threat to its various aspects, when destruction, looting and theft affect parts of it, while oblivion or, at best, change affect the rest. In the midst of Syria's civil war and the stalemate that has followed, preservation of heritage is not always a priority. More essential needs, such as food, medicine, education and shelter, take precedence, and the cost of preserving heritage might seem a luxury that has no place in the battlefields.

However, the close connection of heritage to identity highlights the importance of preserving it and reviving and restoring what has been destroyed by the war. This represents a part of the efforts to recover, assert national identity and even build peace and achieve stability in a society that has been divided by war. This is affirmed by UNESCO, the United National Educational, Scientific and Cultural Organization, in the 1954 Convention for the Protection of Cultural Property in the Event of Armed Conflict, commonly referred to as the Hague Convention. "Cultural heritage is a mirror of the life of a society, its history and identity. Its preservation helps to rebuild societies that are divided, to restore their identity and connect their past with their present and future," UNESCO said in a statement[1].

In March 2011, a civil uprising in Syria began with peaceful protests that soon turned into an armed conflict that, seven years later, has

---

[1] UNESCO, "The Convention for the Protection of Cultural Property in the Event of Armed Conflict" [electronic reference], available on the UNESCO website at http://portal.unesco.org/en/ev.php-URL_ID=13637&URL_DO=DO_TOPIC&URL_SECTION=201.html, 2017, accessed on March 11, 2020.

killed hundreds of thousands, displaced nearly six million Syrians outside the country and displaced more than six million people within Syria. At the economic level, Syria's cumulative losses in gross domestic product between 2011 and 2017 have been estimated at more than $200 billion in US dollars, which is four times the nation's total GDP in 2010.[2] Moreover, 30 percent of the country's infrastructure, homes, and health and educational facilities have been damaged. This figure is estimated at about 23 percent in Damascus and its countryside, which was transformed into violent clash zones between 2012 and 2018[3].

Since 2011, many aspects of Syria's tangible and intangible cultural heritage have been subjected to destruction, vandalism and material and human change. Amid instability and lack of signs of the end of the conflict, it seems that preserving and protecting these assets, and restoring what was destroyed, is a difficult and non-priority task. This highlights the role of various institutions and cultural research in shedding light on the aspects that we can work on to avoid losing these cultural assets as much as possible.

This research stems from a question about the changes that have taken place in Damascus, and whether these crafts have become threatened with extinction due to the ongoing war. It aims to study the current reality of these crafts in terms of labour, raw materials, machinery, production and marketing, and to compare their current condition to their reality before 2011. The research also aims to try to discover the most important difficulties and challenges facing practitioners of these crafts, and to formulate a set of recommendations regarding what should and should not be done to prevent further deterioration. These recommendations were the results of dozens of interviews with workers and specialists in these crafts, conducted during the period of preparation of the research.

---

[2] Harun Onder et al., "The Toll of War: The Economic and Social Consequences of the Conflict in Syria" [electronic reference], Washington, DC: World Bank Group, 2017. Available at http://documents.worldbank.org/curated/en/811541499699386849/full-report.

[3] For more details about the effects of the war on Syria's various sectors, please refer to the overview of the humanitarian needs of Syria for 2018 prepared by the Office for the Coordination of Humanitarian Affairs at the United Nations: https://hno-syria.org.

The importance of this research lies in the significance of the traditional Damascene craftsmanship and its connection to the city's identity in particular, and to Syrian identify in general, and in the serious reality that these crafts could be faced with extinction unless organised and systemic efforts are undertaken to preserve them following the profound changes that have touched their various components. The survival of these crafts does not only mean preserving traditional pieces of cloth produced here or there, but rather preserving their identity, form and characteristics without neglecting any of them, thus preserving a Syrian history threatened by extinction in many of its aspects.

Today, the amount of research available in Syria on the current status of traditional handicrafts is scant. This is especially true of research studying specific handicrafts in detail. We have been able to document an academic thesis published at the University of Damascus in 2016 under the title "The Brocade Fabric Industry and the Significance of Brocade-Related Decorations"[4]. It discusses the brocade industry and the difficulties it faces. In addition, we have found dozens of articles published since 2011 emphasising the importance of protecting the Syrian crafts from extinction. Although we cannot be sure that the 2016 paper and the articles we cite are the only published information in this regard so far, our long search and inability to document any similar research also highlight the importance of this paper in terms of recommendations and results that we aspire to put into the hands of parties capable of implementing what is possible.

This research embraced a descriptive approach. At first, five Damascene fabrics were chosen as examples for the study, namely brocade, the Aghabani textile, handmade rugs, and the Damasco and Sayah fabrics. Information and testimonies were collected on these crafts' situation before and after seven years of war in order to obtain analysable results. After a survey of the current status of the selected Damascene fabrics, case studies were carried out to determine the reality of these crafts to gather a range of data, including, as far as possible, the number of remaining workshops and machines, number of current

---

[4] Salem al-Ahmad Abdullah, "The Brocade Fabric Industry and the Significance of Brocade-Related Decorations", a master's thesis in Rehabilitation and Specialization in Folklore, Damascus University, Faculty of Arts and Humanities, Department of Sociology, Damascus, 2015/2016.

workers, production methods, access to raw materials, and marketing and selling methods.

In order to carry out the case studies, we identified the research community, which consists of practitioners of these crafts at all levels, fields and stages of production, in addition to various shops engaged in the sale of Damascene cloth products, and related government bodies, including the General Union of Syrian Craftsmen, the Union of Craft Associations in Damascus, and the Second Industrial Secondary School. Depending on the nature and size of smaller groups within the larger research community, the size of the sample and the number of community members to be studied were determined. For the smallest groups, comprehensive samples were selected and all members of the group were interviewed, whenever possible, as in the case of the remaining persons who are dyeing the silk yarns used in the brocade industry, or of those who specialise in repairing the machines used for embroidering the Aghabani textiles. As for the larger groups within the research community, which are difficult to survey, we have chosen objective samples that represent the group for consideration, as in the case of souvenir shops selling traditional crafts of the Levant.

Individual interviews were conducted face-to-face with the respondents to ensure full awareness of the subject matter of the study and to integrate the provided information with field observations. We conducted 26 group or individual interviews and asked open questions about the current reality of the textile industry, its future prospects and related recommendations. These interviews can be summarised as follows: fourteen interviews with producers and vendors of different types of Damascene fabrics, five interviews with artisans working in different stages of Damascene fabric production, five interviews with owners of shops selling traditional crafts and textiles as souvenirs, one group interview with a number of workers in the textile department at the Second Industrial Secondary School, and one interview with an official of the Union of Craft Associations. These interviews took place over periods ranging from half an hour to three hours, and some interviews were repeated more than once to get more information upon need. It is worth mentioning that eight of the interviews were conducted with the aim of confirming or refuting some of the information collected and were used neither in quotations nor in direct evidence within the research. This was part of our efforts to develop a comprehensive understanding of the subject of research in all dimensions and increase the level of knowledge, which we have tried to accomplish as much as possible within the temporal and spatial limits of the research. This does not preclude the possibility of mentioning some information that is likely to need expansion and further verification.

After collecting all the information, we decided that the research should be limited to three crafts: brocade, the Aghabani fabric, and handmade rugs. We did not receive any information adding value concerning the Damasco and Sayah fabrics, since most of those crafts' workshops have been located outside Damascus, in Aleppo in particular, since more than ten years before the onset of the war.

In addition, we have used a number of references to complete the theoretical basis of this study. These include a collection of books and research studies that helped us to learn about the history of traditional Syrian crafts, especially the textile handicrafts, in addition to the characteristics of the chosen crafts in terms of their most important properties, the tools and the raw materials used to produce them, and the hierarchy of workers.

Challenges we faced included the difficulty of identifying all members of the research community in some cases, and the oral transmission that is still practised among members of this society in documenting the stages and details of the crafts, some of which still not documented in written form. In addition to the difficulty of obtaining all the information, there was also the challenge of resolving any contradictory information provided by different sources.

At times it was difficult to find a source whom we wanted to interview. For example, we spent a whole day trying to reach a man said to be working in the printing phase of the Aghabani production, but did not manage to find him. Many people knew his name and address in an approximate manner but did not know whether he was still practising his work of even if he was still alive.

At other times a source was reluctant to talk, perhaps out of boredom or a feeling of despair with the current situation. We sought to overcome this challenge by diversifying the sources and interviewing the largest possible number of professionals in the various stages of the studied crafts, in order to get a full picture of the previous and current reality of these crafts, which we hope we have done as much as possible.

In the end, the research comes out with a set of recommendations put forward by the interviewees themselves, and we believe that it could contribute, if applied by the concerned parties, to preserve these Damascene crafts in particular and Syrian crafts in general from the risk of extinction. These recommendations focus on various mechanisms to encourage new workers to learn crafts, to support the marketing of handicrafts internally and externally, and to raise public awareness regarding the significance and value of traditional Syrian crafts.

Part One – Damascene Textile Crafts from Ancient Times until 2011

In this part, we will discuss the Damascene textile crafts in general, highlighting their most important features and characteristics, their legal and regulatory status, and their association with Syria's economy. We will address in detail the three crafts we chose as examples for the study; we will list their most important characteristics, methods of production and their reality before the war in Syria in 2011.

1. A Historical Overview of the Textile Crafts in Syria in General and in Damascus in Particular

The textile industries are among the oldest, most prosperous and most widespread of Syrian handicrafts. Syria has long been known for products that bear the authentic character of societies that flourished in the Levant during successive historical periods. The region had a high position in these industries throughout the ages. The ancient Canaanites were famous for their woollen textiles, and the textile industries flourished during the Roman era and highly improved during the Islamic era, especially during the Umayyad period.

Historical documents dating back thousands of years, including hieroglyphic inscriptions, state that "Syrians were the masters of the textile industry in the late 4th millennium BC and that their textiles were spread throughout the known world"[5]. Remnants of woollen, silk and flaxen fabrics, the sculptures discovered in the tombs of Palmyra, as well as the remains of a princess's dress in the city of Baniyas al-Janub in the province of Quneitra, illustrate the importance of these traditional Syrian handicrafts and the art of sewing, and their aesthetic and long history throughout the ages. These documents and artefacts also

---

[5] Mounir Kayyal, *Levantine Achievements in Damascene Arts and Industries*, a publication of the Syrian General Authority for Books, Ministry of Culture, Damascus, 2007, p. 56.

show the extent of interest in the use of embroidery to confer beauty on the clothes of men and women[6].

The Syrian textile industry has depended, for ages, on several stages that are necessary prior to weaving. The threads required for this process are prepared through successive manual activities, including the processing of yarns and assembling them in blocks called kibabs before winding them onto special pipes or cylinders of equal size, including some used in dyeing the yarns in the preferred colour and rewinding the ones that might be cut upon dyeing them, as well as others related to the placement of these threads on the weaving machine in certain patterns in preparation for weaving. The weaving process itself comes among the late stages of producing a piece of cloth. It is followed by finalization processes, such as washing and ironing, before the product is finished and is ready for sale and trading.

Each of these stages has professionals who master them with great precision as the raw materials or the unfinished pieces move among them according to each professional's area of competence and qualifications[7]. Most of the Syrian handicraft skills are of a family nature; that is, they are inherited. The workmanship passes from the grandparent to the parent and then to the child, so the practitioner becomes the companion of the career early in life, allowing him or her to embrace the spirit of craftsmanship and to see its secrets and mysteries skilfully and flexibly[8].

Damascus has been famous for its textiles since ancient times. Its geographical location on international trade routes has contributed to the popularity of its textiles, characterised by quality, professionalism and low prices. In the Islamic era in particular, the arts and industries of Damascus were characterised by special features that helped the city to take the first place in many such industries.

In this regard, al-Idrisi, a geographer and historian who lived between AD 1099 and 1160, wrote that Damascus in his time was "a hub of craftsmanship, types of industries and silk fabrics, such as the Khazz

---

[6] Bashir Zuhdi, *Studies in Damascene History, Archaeology and Craftsmanship*, Dar Al-Hilal Publishing House and Dar Al-Yanabi' Publishing House, Damascus, 2010, p. 128

[7] Mounir Kayyal, *Levantine Achievements in Damascene Arts and Industries*, p. 6.

[8] Mounir Kayyal, *Levantine Achievements in Damascene Arts and Industries*, p. 34.

and precious, exquisitely made one-of-a-kind Dibij brocade[9], which are carried from it [Damascus] to every country and taken from Damascus to all the horizons and nearby countries, and those located far away. Its factories in all these fields were wonderful, imitating the beauty of the Byzantine silk Dibij, and are equal to the fabrics of Tustar[10] and Isfahan[11], and the Tiraz of Nishapur[12], and the glorious designed silk clothes, and the clothes sold in Tanis[13]. It contains precious clothes and great beauties that have no equivalent and nothing else can approach them"[14].

Abu al-Baqa' al-Badri, a writer and poet who lived in the 15th century AD, also wrote: "Of the beauties of Damascus is what is manufactured there of cloths of various types, engravings and drawings, such as the Atlas silk[15], the Hormuzi cloth[16] and silk fabrics"[17].

Many factors led to the spread of the Damascene textile crafts, including their originality at the global level, the possibility of preparing the raw materials in the industrial centres on favourable economic terms, the availability of experienced skilled craftsmen who inherited their workmanship from their forefathers, the presence of flourishing internal and external markets able to absorb the products, and the

---

[9] According to al-Mu'jam al-Waseet dictionary, Khazz is a cloth woven of wool and silk, while Dibij is a fabric of authentic silk of different colours.

[10] Tustar is a town in the Ahwaz region in southwestern Iran that was famous for its textile industry and trade.

[11] Isfahan is an Iranian city located 340 kilometres south of Tehran, in the centre of Isfahan province. It is famous for its heritage and large organised markets. It is on the UNESCO World Heritage List.

[12] Nishapur is a city in Razavi Khorasan province in north-eastern Iran. It is one of the oldest cities in Iran. It was famous for its culture and architecture, especially during the Abbasid period.

[13] Tanis is an Egyptian city located near Damietta and well-known since the pre-Islamic era for the manufacture of high-quality textiles and clothing.

[14] Mounir Kayyal, *Levantine Achievements in Damascene Arts and Industries*, p. 23.

[15] The Atlas fabric is a textile characterised by its lustre, draping and softness due to the low number of intersecting points of the longitudinal and lateral lines, or "warp and weft."

[16] Hormuzi is a high-quality coloured silk fabric which is particularly famous in the women's clothing industry in the Levant.

[17] Abu al-Baqa' al-Badri, *Nuzhat al-anam fi mahasin al-Sham (A Night Walk in the Beauties of the Levant)* [electronic reference], available at https://goo.gl/aVkDqF, [n.d.], accessed on March 7, 2018, p. 362.

ability to secure good material suppliers for these crafts and the industries' workers[18]. This spread and the internal and external demand for procurement have highlighted the need to gather such industries in close proximity; thus the workshops specialised in manufacturing a single type of crafts gathered in some neighbourhoods of Damascus and its countryside. For example, the Qaimaria[19] neighbourhood was called Little India, where the textile industry and trade was concentrated. Other specialised areas include al-Shaghour[20] neighbourhood, al-Tariq al-Mustaqim[21] ("the straight road"), which includes the markets of Medhat Pasha[22] and Bab Sharqi[23], and other Damascene neighbourhoods and markets[24].

As a craft is generally an integrated effort between labour, machinery and raw materials, craftsmanship has to be organised within regulatory frameworks concerned with production and marketing and able to solve potential problems between artisans. Thus, craft organizations were established in the Ottoman era and played a significant role in economic and social life. The traditional industries are each governed by certain customs and traditions. The most important of these is the Sheikh of the Craft (Sheikh al-Kar) and the system of varieties ("asnaf")[25], where each sheikh is concerned with a particular occupation or a specific craft in the field of production, service and marketing, all of which follow a similar sequence of five orders and two equivalent ranks[26]:

---

[18] Mounir Kayyal, *Levantine Achievements in Damascene Arts and Industries*, p. 31.
[19] Qaimaria: An old neighbourhood of Damascus with many monuments and houses built in the old Damascene architectural style.
[20] Al-Shaghour: One of the oldest neighbourhoods of Old Damascus.
[21] Al-Tariq al-Mustagim Street: A road located in the heart of Old Damascus, extending from its west to the east. It was built in the first century BC.
[22] A Damascene market extends from Bab al-Jabiyah to the Roman Triumphal Arch. Some of its parts are covered with iron roofs, like many of the old Damascene markets.
[23] Bab Sharqi, a neighbourhood located on the eastern side of Old Damascus. It is named after Bab Sharqi ("the Eastern Gate"), which is one of the gates of Damascus' old wall. It has many shops and workshops to manufacture many traditional Damascene crafts products.
[24] Mounir Kayyal, *Levantine Achievements in Damascene Arts and Industries*, pp. 31–36.
[25] Mounir Kayyal, Levantine Achievements in Damascene Arts and Industries, pp. 347–348.
[26] Mounir Kayyal, *Levantine Achievements in Damascene Arts and Industries*, p. 348.

1) The Sheikh of Sheikhs: This rank comes at the top of the crafts pyramid and symbolically oversees the craft and in particular the inauguration of the sheikhs of crafts. The Sheikh of Sheikhs helps the captain appointed by him or her, who acts as the sheikh's representative in the meetings of the craft organizations and is required to carry out the duties given by the Sheikh of Sheikhs.
2) The Sheikh of the Craft (Sheikh al-Kar): The head of the profession, a position granted for life, is chosen by the Mu'allims, or "teacher craftsmen," and can be dismissed at their request or upon the request of the Sheikh of Sheikhs. He must be a meticulous professional, with high ethics, and an honest and wealthy person, for he has many responsibilities. He is considered the reference in determining the customs of the craft. The Sheikh of the Craft helps the Shawish, who is appointed by him to carry out every task assigned to him.
3) The Mu'allim ("Teacher Craftsperson"): These are the heads of a particular craft in their own establishment and are proficient in the craft. They are usually independent merchants employing artisans and apprentices.
4) The Artisan: A worker in the establishment of a mu'allim who works in the craft but is not as skilful as his or her teacher.
5) The Apprentice: Usually a child who works without a wage under the tutelage of an artisan until reaching adulthood.

In the course of the research, we identified certain proverbs and folk songs related to the textile industry (see Appendix 1 for a number of examples). We have also noted that the surnames of some Syrian families, especially those from Damascus, signify specific professions, including professions related to the textile industries. Examples of family names in this category include "Hayek," "Dakkak," "Habbal," "al-Sabbagh," "Noelati", "Musallati" and "al-Sawwaf," among others. These details are only a few among other indications of the extent to which these crafts relate to Syrian identity in general and to Damascene identity in particular.

In this section, we provided an overview of the history of the textile crafts in Syria and Damascus, which highlights the heritage of these crafts, their originality and their close association to the country's identity. In the following section, we will give an overview of the legal and regulatory situation of the Syrian crafts and their association with the Syrian economy.

2. An Overview of the Regulations and Legal Reality of the Syrian Crafts

"Herfa," the Arabic word for a craft, is derived from "ihtiraf," or professionalism. It means working in a certain craftsmanship, based on personal effort and integrated skills, to produce a unique, distinct and artistic product. The more formal definition of a craftsman is someone who works in producing materials and providing professional services based on personal effort and professional experience with the help of other workers[27].

In Syria, artisans account for 8 percent of the total population, and 60 percent of them are between the ages of 20 and 44. Since the mid-20th century, Syrian artisans have been involved in trade unions organised by the Ministry of Labour and Social Affairs in terms of their legal regulation[28]. In 1969, the General Union of Syrian Craftsmen[29] was established by Legislative Decree No. 250, which defined the union as "a popular organization that consists of Syrian Arabs and other craftsmen and producers of services. It aims to involve artisans in building the society by organising production and services and working to increase and improve them, enhance their quality, and take care of the physical, moral, health, cultural and social interests of the craftsmen and raise their living standards"[30]. This decree also defines craftsmen as "those who work in the production of materials or those who provide

---

[27] Mohammed Fayyad al-Fayyad and Majed Hashim Hammoud, *Traditional Crafts in Syria*, first edition, translated by Majd Hamoud, General Union of Craftsmen, Office of Culture and Media, Damascus, 2011, pp. 387, 388.

[28] Mohammed Fayyad al-Fayyad and Majed Hashim Hammoud, *Traditional Crafts in Syria*, pp. 389, 390, 413.

[29] The General Union of Syrian Craftsmen's website: http://www.gfcrafts.com.

[30] The General Union of Craftsmen page [electronic reference], on the economic website, at https://goo.gl/wWdbnX, [n.d.], accessed on July 5, 2018.

professional services based on his/her personal effort and professional experience, mainly with the help of family members or other workers, provided that the number of workers in the enterprise does not exceed nine workers"[31].

According to the above-mentioned decree, Syrian artisans were organised within professional crafts associations which exist alongside affiliated institutes and vocational education schools. These associations supervise the training of artisans in various fields and provide them with facilities such as raw materials and a financing process, enabling them to open workshops to practise their craft in. These projects are considered essential elements of economic development in the country.

The associations also contribute in the marketing of artisanal products inside and outside Syria. Before 2011, the percentage of the products marketed within Syria was estimated at 65 percent, while the rest was exported to other countries. Internally, artisans rely on self-marketing or cooperating with various entities to display and sell their craft products, including specialised centres in the archaeological markets and official exhibitions held periodically. In Syria, artisans rarely rely on intermediaries to promote their goods. They prefer to maintain a direct relationship with customers, while they may deal with merchants and shop owners to sell their products. Outside Syria, Arab, European, American and Australian markets are considered the most important markets for Syrian artisanal products[32].

As for the traditional Syrian craftsmanship, which is an important part of the Syrian cultural heritage, the Ministries of Culture and Tourism have supported it by linking the industry to the archaeological sites and preserving them, for they are considered the incubators of such crafts[33]. The Ministry of Tourism has also held periodic meetings and workshops with stakeholders in this regard, headed by the Gen-

---

[31] Mohammed Fayyad al-Fayyad and Majed Hashim Hammoud, *Traditional Crafts in Syria*, p. 388.
[32] Mohammed Fayyad al-Fayyad and Majed Hashim Hammoud, *Traditional Crafts in Syria*, pp. 406–408.
[33] Mohammed Fayyad al-Fayyad and Majed Hashim Hammoud, *Traditional Crafts in Syria*, p. 409.

eral Union of Craftsmen and the Ministry of the Economy, to find the best ways to promote traditional industries[34].

In addition, the Syrian Ministry of Industry supports the Small Enterprises Development Commission[35], which provides services to entrepreneurs and small-business owners, and establishes clusters and incubators for such projects, including projects on traditional crafts. The Union of Syrian Exporters contributes to the marketing of handicraft products through the establishment of internal fairs, the promotion of exporting these crafts by participating in external fairs, facilitating export, and also through electronic online promotion.

Syrian craftsmen are given the option of joining the General Union of Craftsmen. In case of choosing this, they are charged an annual fee of approximately 10,000 Syrian pounds (worth about $23 in US dollars)[36]. In return, they are supported by the union at several levels, such as the ability to participate in exhibitions and access to facilities related to acquiring raw materials, finance and support for the purchase and marketing of products. In addition, they get access to the Social Assistance Fund, which provides financial assistance in situations such as childbirth, death, disability, work injuries, etc., as well as a health insurance fund.

In 2011, the number of professionals working in Syrian traditional industries hit 161,000 people, and these industries contributed 6 percent of the nation's gross domestic product[37]. Early in 2011, the number of craftsmen practising traditional Levantine industries in Damascus was 400, 250 of whom were members of the Union of Craft Associations in Damascus, and 150 of whom were outside the union. The last number is approximate, according to the estimates of the union[38]. The

---

[34] "Traditional Industries in Syria: A Confirmation of Our Civilised Identity" [electronic reference], on *Al-Fidaa* newspaper website at https://goo.gl/d3rVAz, published on December 8, 2011, accessed on January 10, 2019.

[35] The Commission's establishment decree can be found on the SANA website at http://www.sana.sy/?p=321434, and can be accessed at http://www.sme.gov.sy/.

[36] The membership fee has increased slightly since 2011 until the current year.

[37] Fadi al-Aloush, "The Ministry of Tourism Announces Support for Handicrafts through a Specialised Directorate" [electronic reference], on the site of Syria Steps, at https://goo.gl/Sw71Dm, published on December 1, 2011, accessed on July 1, 2018.

[38] Zeina Shahla (the author), "The Reality of the Damascene Crafts Before and After the War", an interview with Khaldoun al-Masuti, of the Damascus Union of Craft Associations, [n.p.], Damascus, on June 21, 2018.

hierarchy of the ranks of these craftsmen was limited to three levels: the sheikh of the profession, artisans and apprentice [39].

The importance of traditional handicrafts in Syria and the governmental attention the sector received did not prevent it from deteriorating gradually in the years before 2011. Many articles and interviews document this decline, which occurred for many reasons. Chief among them was a reluctance to learn these trades due to their difficulty and the low financial returns they offer, compared to other professions such as those related to trade or modern technology, as well as a lack of demand for the products of these crafts, which is limited to tourists and some government bodies and official delegations [40].

In the following paragraphs, we will discuss the three crafts we chose in this study as a model of the Damascus heritage crafts and their most important features before the outbreak of the war in Syria in

---

[39] According to the above-mentioned interview with Khaldoun al-Masuti, the craftsman is a person who owns a legal license and has workers and raw materials. The sheikh of the profession is a craftsman who knows all the secrets of his work and the raw materials needed for it. He has a good reputation and legal status, and is able to innovate and not just to make copies in his craft.

[40] See the following articles on the deterioration of some of the traditional Syrian crafts at various levels before the war:
"Heritage Products Struggle against Time to Preserve the Identity of Communities" [electronic reference], on the Naba' website, at https://goo.gl/gTD58B, published on October 5, 2010, accessed on July 5, 2018.
Y. Ibrahim, "As Most of Handicrafts Have Become Threatened by Extinction, They Are Attracting Visitors to Tartus Summer Festivals 2010" [electronic reference], on *al-Wehda* newspaper website, at https://goo.gl/Xinqya, published on August 8, 2010, accessed on July 5, 2018.
"Including the Brocade Fabrics, the Aghabani and Ajami, Damascene Handicrafts Are in Danger of Extinction" [electronic reference], on *al-Wehda* newspaper website, at https://goo.gl/owwAoT, published on January 31, 2011, accessed on July 5, 2018.
Osama Makiyya, "Abu Mahmoud: The Last Silk Dealer in Damascus and the Profession of Handmade Glass Is Dying" [electronic reference], on Syria News website, at https://goo.gl/wsHqm7, published on March 26, 2010, accessed on July 5, 2018.
Hassan Salman, "The Silk Industry in Syria from Prosperity to Decline" [electronic reference], on Al-Bayan website, at https://goo.gl/enZbH1, published on October 2, 2009, accessed on July 5, 2018.
"Handicrafts Demonstrate the Skills of Syrian Craftsmen" [electronic reference], on D Press website, at https://goo.gl/P9iHJo, published on June 2, 2010, accessed on December 10, 2019.

2011, which will allow us to compare their status then with their situation seven years later.

3. An Overview of the Damascene Textile Crafts Under Study

After reviewing references and conducting interviews, we chose to focus on three Damascene textile crafts, namely brocade, the Aghabani, and handmade rugs, and to study their current situation under the shadow of war and compare that to how things were before 2011. These three crafts were chosen for their great connection to the Damascene identity, as well as the great changes that occurred in them during the war. The following is a summary of each of them:

A) Brocade

The brocade industry started in Damascus several centuries ago[41]. Brocade is a woven silk fabric that consists of fine silk threads and golden or silver synthetic-fibre threads.

---

[41] There is a dispute about the date of the beginning of the brocade industry in Damascus among the different references. Some mention that it dates back to five centuries ago; others say that it dates back centuries but give no precise identification, and still others place it back to 200 or 100 years ago, but not more. It is most likely that the silk industry in Syria dates back hundreds of centuries because of the city's location on the Silk Road, but the brocade fabric as we know it now began about 200 years ago only.

Photo 1: A Brocade selling shop at al-Takiyya al-Sulaymaniyah - Damascus - Zeina Shahla - 2018

There are different narratives to explain the origin of brocade's name. The most common narrative is that "brocade" is derived from "bro," which is a nickname of Ibrahim, and "kar," which means the profession of ancient origin in Kurdish. Other accounts say it has come from the French word "brocart," which means a brocaded or embroidered cloth, or from the Italian word "brocatello," which means a silk cloth embroidered with golden or silver threads[42].

There are different kinds of brocade according to the type of silk used. Its colours vary according to demand and range from two up to seven colours. The designs woven into the fabric also vary from anthropomorphic to botanical and zoomorphic motifs, or simple general designs such as butterflies, nutshells, Saladin motifs and the Damascene jasmines and roses. Brocade is used in the manufacture of wom-

---

[42] We have relied on the following articles regarding the origin of the brocade name:
Alaa al-Jammal, "Sherko Matini: The Brocade, A Damascene Beauty with a Folk Heritage Imprint" [electronic reference], on eSyria website, at https://goo.gl/65KwL3, [n.d.], accessed on July 5, 2018. Joseph Zeitoun, "The Damascene Brocade" [electronic reference], on Joseph Zitoun's website, at: https://goo.gl/KVkPKQ, published on February 26, 2016, accessed on July 5, 2018.
Satik Sinan, "The Damascene Brocade: The Queens' Cloth Industry Is Endangered" [electronic reference], on the *New Arab* website, at https://goo.gl/sQbKBH, published on August 16, 2015, accessed on July 5, 2018.

en's clothing and accessories, some men's clothing, and also some house interior decoration.

Brocade production is a complicated process based on precise work. The motif is drawn on millimetre grid paper and applied to cardboard plates, called Jacquard[43] cards, which are placed in a device on top of a loom. The loom contains thousands of threads and needles that move according to the holes punched in these plates to produce the desired motif. The worker cannot produce more than 10 centimetres of brocade fabric per hour. Thus, brocade production depends largely on demand and touristic activity.

Photo 2: Jacquard cards placed at the top of a loom – al-Takiyya al-Sulaymaniyah, Damascus - Zeina Shahla - 2018

Of what has been written regarding the Damascene brocade in Western-oriented newspapers in the 20th century, we read: "Brocade is a victory achieved by the Syrian craftsmen, a marvel of the wonders of

---

[43] The Jacquard cards are cards made of a special thick cardboard with holes punched in specific locations and used to store information that can be read mechanically. Cards of this type date back to the Middle Ages and were first used in textile machinery by the French merchant and weaver Joseph Marie Jacquard in the early 1800s.

Syrian industries," and "Brocade is a high-end industry that cannot be matched or imitated"[44].

The brocade industry is divided into two types: manual and automated. The manual craft requires difficult hand and foot work on traditional looms, while the automated craft uses modern power looms to carry out the work of weaving and requires the craftsman's oversight only. Experts can distinguish between the pieces of brocade fabric produced manually and those produced on mechanical looms, as the latter have less aesthetic appeal due to a slight increase in thickness and texture of the yarns used, because of the power of the automatic shuttle. While machinery-produced brocade started in the 20th century, the manual looms inside Damascus continued to operate at a decreasing pace, until they were completely halted after the outbreak of the current war[45]. In 2010, a newspaper article documented the existence of three manual looms in use to produce brocade with fewer than three mu'allims ("teachers") and no more than ten workers[46].

Photo 3: A manual brocade loom – al-Takiyya al-Sulaymaniya - Damascus - Zeina Shahla – 2018

---

[44] Kamal al-Qasimi and Hasan al-Hamami, *A Report on Traditional and Artistic Industries and Handicrafts in Syria*, [n.p.], 1975, p. 2.

[45] Zeina Shahla (author), "The Reality of Brocade Craftsmanship", an interview with Mohammed al-Rankoussi, Damascus, [n.p.], on January 11, 2018.

[46] Ammar Abu Abed, "Damascene Brocade Weaves Silk with Silver and Gold Threads" [electronic reference], on the website of *al-Ittihad* newspaper, at https://goo.gl/jBbmsu, published on March 9, 2010, accessed on July 5, 2018.

The most famous factories producing machine-made brocade that continued to operate in Damascus and its countryside up until the onset of the war were: Mouzannar in Adra Industrial City[47], al-Asil in Qaboun, al-Rushi in Beit Sahem, Mteini in Jaramana, and Na'asan in Bab Sharqi, among others. These plants had dozens of looms, and it was difficult to determine their monthly or annual production capacity. They were completely dependent on demand in the market[48]. Before the outbreak of the war in Syria, brocade vendors were largely dependent on tourists who wanted to buy these fabrics for their beauty and for various uses such as gifts or to decorate houses and palaces. Various governmental institutions were among the most important customers for brocade, purchasing it for the purpose of giving it to the official and diplomatic delegations that visited Syria throughout the year.

As for the raw materials, brocade production used to depend on local silk yarns from several factories, most notably a factory in the Duraykish area, near Tartus, in western Syria. But silk production in Syria has gradually declined since the early 1990s[49] and has been replaced

---

[47] Antoun Mouzannar is one of the foremost craftsmen in the brocade industry in Damascus. His grandfather, also named Antoun Mouzannar, founded the first textile factory in the Damascus region in 1890. The Mouzannar company is currently located in Adra Industrial City. See Shahla Zeina (author), "The Reality of Brocade Craftsmanship", an interview with Antoun Mouzannar, Damascus, al-Harika, [n.p.], March 18, 2018.

[48] The information in this paragraph is based on the interviews we conducted with brocade producers and which are documented in the References section.

[49] We have relied on the following articles regarding the decline in local production of natural silk since the early 1990s:
Hisham Adra, "Studies Confirm the Feasibility of Reviving the Syrian Silk Industry" [electronic reference], on *Asharq al-Awsat* website, at https://goo.gl/FRfoKN, published on April 8, 2001, accessed on July 3, 2018.
Fatina Abbas, "Sericulture in Tartus: a Long-Standing Career on the Road to Demise" [electronic reference], on the Panorama Tartus website, at https://goo.gl/djBjM7, published on May 1, 2016, accessed on July 3, 2018.
Yahya Mohammed Haitham, "The Government Delegation Headed by Engineer Khamis Continues his Field Visit to Tartus Province" [electronic reference], on *Al-Thawra* newspaper website, at https://goo.gl/brjwG9, published on April 17, 2017, accessed on July 5, 2018.
Ghosun Deeb, "Silkworm Breeding in Tartus: The Lack of Marketing and the High Cost of Supplies Are the Most Important Reasons behind the Decline in Production"

by imported Chinese silk, in addition to synthetic-fibre threads imported from various sources, mainly Germany, France and China[50].

The production of brocade involves several stages, some of which do not require specific skills, while others require techniques the worker must thoroughly master, taking into account the fragility of the silk yarns used in the industry.

The following are the stages of brocade production in order, with a simplified explanation of each:

1) Silk Twisting: This process includes kinking or wrapping silk threads onto wooden rollers in a precise, accurate and regular manner. This stage needs equipment such as rollers and wheels, and is no longer a very tiring process for craftsmen, as they get the imported silk already twisted.

2) Dyeing: This includes removing a gum of sticky protein from the silk fibres and dyeing the silk in the required colours.

3) Silk Spinning: This means the winding of silk after dyeing onto spools using a special machine called a spinning machine.

4) Silk Warping: This means the placement of silk threads onto the loom in a longitudinal direction (the warp), with the number of threads ranging between 4,000 and 10,000 according to the motif of the product to be woven. People usually learn this skill by having it passed down from their parents or at factories by observing those who have mastered this stage.

5) Threading: This is the linking of two threads of different colours in the warp, a stage found in most textile industries and not only brocade. Special machines can be used if needed.

6) Punching the Jacquard cards: This process has two types—a manual one, which has completely fallen out of use for more than ten years, and a computerised type, still

---

[electronic reference], on *Al-Thawra* newspaper website, at https://goo.gl/qwdKK7, published on January 14, 2012, accessed on July 5, 2018.

[50] This information was obtained through interviews with brocade producers in Damascus.

used by a limited number of people working in Damascus and Aleppo.
7) Weaving/Knitting: The craftsman weaves the cloth according to the longitudinal warp threads and the previously set shuttles which carry the horizontal threads (the weft), a stage requiring great craftsmanship and proficiency. A department to teach this skill at the Second Industrial Secondary School in Damascus was opened in 1994.
8) Softening (Compression): This means ironing the cloth and compressing it so it becomes smooth using special machines that are available in several workshops in Damascus.

B) Al-Aghabani

Al-Aghabani is well-known as a textile used for interior decor items such as tablecloths and tea sets, as well as headscarves and wedding dresses. The craft began in Aleppo before moving to Damascus. According to one account, the name "Aghabani" refers to the orange-yellow colour of saffron; another account attributes it to the cruciform shape of the poles used in embroidery[51]. A third account relates that the origin of the word refers to the first two families who produced the Aghabani, which began as a partnership between two members of the Aga family and the Albani family. Thus the word al-Aghabani resulted from merging the names of the two families[52].

The Aghabani is made using cloth of cotton or linen, which is embroidered using silk and synthetic-fibre threads. The patterns of the embroidery consist of botanical and Arabian motifs, such as roses, flowers and twigs. These motifs are engraved on wooden templates and then printed on the fabric as many times as required with various techniques, such as spraying and effacing. As for the engravings, they have several designs, most notably the scabiosa (pincushion flower),

---

[51] *The Intangible Syrian Cultural Heritage: Skills Associated with Traditional Craftsmanship, Part I,* Damascus: Syrian Ministry of Culture and the Syria Trust for Development, 2014, p. 61.

[52] Zeina Shahla (author), "The Reality of the Aghabani Craftsmanship", an interview with Safouh al-Mawli, Damascus, [n.p.], on December 30, 2017.

olive, jasmine, pine nuts, vine leaves, chrysanthemum and the hall-ceiling motifs[53].

Photo 4: The Aghabani cloth - Damascus - Maher Al Mounes – 2018

The Aghabani production includes several stages, with groups of women and men working at each stage in different places. Among these stages are printing the motifs on the cloth using wooden moulds and special ink, embroidery, and processing the final piece. Men often work in the printing and processing stages, while women—most of whom are residents of the Damascus countryside, especially the Eastern Ghouta region—specialise in embroidery. The embroidery originally was done by hand before it developed into a semi-automatic and semi-manual process using specialised sewing machines[54].

---

[53] In this paragraph, we have relied on the following resources:
The Intangible Syrian Cultural Heritage: Skills Associated with Traditional Craftsmanship, Part I, Damascus: Syrian Ministry of Culture and the Syria Trust for Development, 2014, p. 61.
Mounir Kayyal, *Levantine Achievements in Damascene Arts and Industries*, p. 80, 81.
Mohammed Fayyad al-Fayyad and Majed Hashim Hammoud, *Traditional Crafts in Syria*, p. 317.

[54] We obtained this information from several interviews with the Aghabani shop-owners and salesmen and other cloth-selling shops in Damascus. The interviews are documented in the References section.

Before the outbreak of the war in Syria, the Aghabani trade was run by Damascus-based tradesmen who knew all the stages of the Aghabani production, including the cloth processing, printing, embroidering, and final washing and ironing. They were the link between the workers in those stages and the traders who bought the final product and sold it in the markets. The stages of processing and printing designs on the fabrics were done mainly by male workers in Damascus who have mastered the skill. The stage of embroidering the Aghabani motifs was mainly concentrated in Eastern Ghouta, especially in the town of Douma and its surroundings (namely al-Shifouniya, al-Rayhan, Autaya, Beit Nayim and Mayda'a), where hundreds of women used to work in it at their homes. It was rare to find a house in Douma where there was not at least one Aghabani machine that was used by most of its women of all generations. This profession was a guarantee for women to learn a craft that could be practised at home with a reasonable material income. "I used to go to Douma every week and there was a large number of female workers who were coming to give me the products and get paid," said one trader. "The queue was like a bread queue in front of bakeries"[55].

Photo 5: Printed Aghabani fabrics before embroidery - Damascus - Zeina Shahla – 2018

---

[55] Zeina Shahla (author), "The Reality of the Damascene Textile Crafts", an interview with Samer al-Nokta, Damascus, [n.p.], on January 11, 2018.

The embroidery stage follows the stage of cloth processing, which is an easy process that does not require any special skill, and the stage of printing the designs that will be embroidered on the cloth. This printing is done with wooden blocks engraved with the desired pattern, wet with a certain ink (indigo dye), and used to stamp the motifs on the cloth with a certain repetition, as a guide for the embroidery. After the embroidery, the traces of ink are removed by washing and ironing the cloth, which are the last stages of the work.

Before 2011, al-Aghabani was based on locally manufactured fabrics mainly from al-Khomassiya[56], al-Debs[57] and al-Maghazil[58] companies, as well as imported embroidery threads of French or Chinese origin[59]. The tools used for the production of the Aghabani are of two main types: the specialised sewing machines used for the embroidery, and the wooden templates used for printing the designs on the cloth.

Ahmed al-Sheikh, one of the workers in the sale and maintenance of the Aghabani machines in al-Shaghour's Bab al-Hadid neighbourhood, estimated that there were about 15,000 Aghabani machines in Douma and its surroundings before 2011, where the Aghabani machine was an essential part of the bride's trousseau. It was called al-oqda, or "the knot," and it could be sold in time of need, just like gold and jewels, as it carried a value of up to $1,000. "My grandmother brought up her five daughters and got them married thanks to the Aghabani machine," al-Sheikh said. "I opened my eyes as a kid while hearing the voice of this machine. The Aghabani was and still is an integral part of our lives, culture and heritage."

Most of the Aghabani machines were of French, German, Belgian, American, Indian or Japanese origin, dating back more than 60 years.

---

[56] The al-Khomassiya United Industrial Trading Company is a Syrian textile company. It was founded in 1946 and nationalised in 1961. It is located in the Qaboun area of Damascus, next to the Bolman garages. It contained hundreds of machines and looms and about 1,700 workers in 2010.

[57] United Arab Industrial Corporation (Debs) is a Syrian textile company founded in 1954 and nationalised in 1961. It is located in al-Qadam area of Damascus. In 2012, it had a total number of about 800 workers.

[58] The General Company for Spinning and Woven Industries is a Syrian textile company founded in 1937 and nationalised in 1961. It is located in the Jubar neighbourhood of Damascus. In 2010, there were about 900 workers working there.

[59] This information was obtained from the Aghabani workers in Damascus.

They are similar in their workings and the types of stitches they produce. However, some machines come with one, two or three needles. The town of Douma was also home to about 10 people who were repairing the machines, and one of them was also manufacturing it locally[60].

Photo 6: The Aghabani printing template - Damascus - Maher Al Mounes – 2018

The Aghabani tools also include the wooden moulds used for printing designs on the cloths. These moulds are made of willow wood and great skill is required to carve them into the desired pattern. It is estimated that each mould can be used for up to two years before it becomes unusable.

C) Handmade Rugs

The word for rugs in Arabic is "beset," the plural of "bisat," or a mat, and is derived from the verb for spreading something out. Many Syrian cities, including Damascus, were famous for the manufacture of handmade rugs, benefiting from the availability of the necessary raw materials, most notably wool, animal hair and fur. The traditional handicraft industry was characterised by quality, beauty and mastery. The

---

[60] Zeina Shahla (author), "The Reality of the Aghabani Craftsmanship", an interview with Ahmed al-Sheikh, Damascus, [n.p.], on April 28, 2018.

rugs were used to cover floors and walls, as many of the rugs' designs were used to decorate Damascus houses. The drawings and decorations used were various as well, including geometrical and landscape themes.

Photo 7: A shop selling handmade rugs in Medhat Pasha - Damascus - Zeina Shahla – 2018

Syria's handmade rugs industry began and developed from the 18th century inside houses, especially in rural areas, where the craft was considered a basic source of income for many families benefiting from the presence of high-quality raw materials. Over the years, this handicraft industry has not disappeared and has preserved its form, design and spirit until the present time. As for the Damascene rug, its fame comes from its quality, craftsmanship, rich colours, and various motifs. The themes of handmade rugs differ from one country to another and from one city to another. Experts can distinguish the rugs made in Damascus from those made in Hama or Homs, and between the Syrian ones and their Palestinian counterparts.

Wooden looms are used in the manufacture of rugs. The loom is made up of several parts, such as the poles that determine the distance between yarns, posts around which threads are twisted, thread sorting devices, stitching tools and staples. The raw materials used in making rugs are wool, cotton, silk, lint, hair, linen and hemp, most of which are locally produced.

The Damascene Textile Industry in the Cauldron of War

Spinners, dyers, painters and weavers work in the manufacture of rugs and carpets. For some of these processes, like spinning and dyeing the yarn, machines have replaced hand work today.

Photo 8: Handmade rugs loom – al-Takiyya al-Sulaymaniyah - Damascus - Zeina Shahla - 2018

The handmade rugs industry in Damascus suffered a sharp decline in the number of workers before the war. In 2009[61], there were five registered workers, compared to 37 looms and more than 300 workers in the middle of the 20th century[62]. This decline is associated primarily with the advent of the quicker and easier-to-use automatic looms.

In this section, we have talked about the Syrian crafts in general, their history, and their organizational and legal status, in addition to reviewing the crafts we chose to study in this research along with their characteristics, features and pre-war reality. In the next section we will talk about the reality of these crafts about seven years after the start of the war, relying mainly on the interviews we conducted during the research period.

---

[61] Hisham Adra, "Five Rug-Producing Manual Looms Are Left in Damascus" [electronic reference], on *Asharq al-Awsat* website, at: https://goo.gl/GmUWJh, published on February 13, 2009, accessed on July 5, 2018.

[62] Kamal al-Qasimi and Hasan al-Hamami, *A Report on Traditional and Artistic Industries and Handicrafts in Syria*, p. 5.

Part Two – The Reality of the Damascene Fabrics in the Shadow of War

Historically, many of the periods of turbulence, crises and wars that have befallen Damascus and the Levant, such as the invasion of Tamerlane[63], the Crusades[64] and the 20th century's wars[65], directly affected the traditional Damascene handicrafts, especially in terms of production, sales and prices. These conflicts had a great impact on the labour force, resulting in a scarcity of skilled workers and the accumulation of goods in stores due to the lack of demand as a result of the region's instability, the cessation of tourism and a lack of demand in foreign markets[66].

Today, we cannot pay a visit to a shop selling Damascene fabrics or traditional crafts souvenirs in the Damascus markets without finding sellers complaining that sales have almost vanished. Shop owners estimate their sales decline after 2011 at 80 to 95 percent, according to the interviews we conducted in this study. Some merchants spend the day sitting in front of their shop playing backgammon or talking to colleagues. In the afternoon, others prefer to turn off the lights and go to sleep, for who, they ask, is that customer who will come to buy these high-priced Damascene antiques, which are a luxury today and come after dozens of priorities in the hierarchy of Syrian families' needs? Where are the tourists who once filled the alleys of the Old City and its markets?

With the widespread destruction of most of Syria during the war, handicrafts had to be negatively affected. This impact fell on several aspects of the handicrafts industry, including the labour force, raw ma-

---

[63] The Mongol invasion of Syria led by Tamerlane that resulted in the destruction of its cities and the capture of Damascus in AD 1400-1401.
[64] A series of campaigns launched by European countries on the Arab region and the Holy Land during the 11th, 12th and 13th centuries AD.
[65] The series of 20th century conflicts affecting Syria, foremost among which are the Tripartite Aggression on Egypt (the Suez Crisis), the civil war in Lebanon, and the wars against the Israeli occupation of Palestine.
[66] Mounir Kayyal, *Levantine Achievements in Damascene Arts and Industries*, pp. 25, 26, 27.

terials, machinery, lines of production, and the sales market. Generally, most of the interviewees described the situation today as catastrophic, a reality that varies from one craft to another according to its nature, the location of its workshops, factories, workers, and the raw materials used in it, as well as the potential final consumer.

There are many reasons for that, varying from the separation of Damascus from its countryside, which is considered the most important reservoir for factories, labour and raw materials, to the depletion of the labour force in general for various reasons, including workers migrating abroad, or engaging in hostilities and suffering disabling injuries, as well as the decrease in the number of tourists who used to visit Syria and buy large quantities of the products of these trades[67]. Other considerations include the weak marketing strategies and the lack of local exhibitions that previously were an opportunity to sell these products, and the difficulty of securing the raw materials necessary for the continuation of work as a result of the exit of many local factories from service because of their location in dangerous or inaccessible areas, as well as the difficulty of importing. All of these factors resulted, as a final outcome, in higher production costs and thus higher final selling prices[68].

This led to a decline in the production and sale of Damascene textile crafts, and the reluctance of many workers to learn or continue to practise these crafts, choosing instead to practise other professions that ensure greater financial stability, or to migrate abroad. These losses are all the more significant if we take into account the familial nature of these occupations, which depend on the inheritance of skills

---

[67] "Minister of Tourism: Tourism Losses 387 Billion Syrian Pounds and a Decrease of 98 Percent in the Number of Tourists" [electronic reference], on Syria News, at https://goo.gl/f2QCq5, published on March 23, 2015, accessed on July 5, 2018.

[68] We have relied on the following articles illustrating some of the war's effects on the Damascene handicrafts in Syria:
Sami Issa and Markazan al-Khalil, "The Reality of the Textile Industry under the Microscope of Experts: The Vision of the Ministry to Provide Financial Support for Rehabilitating the Expert Hand" [electronic reference], on *Tishreen* newspaper website, at https://goo.gl/Sj4bjm, published on July 25, 2016, accessed on July 5, 2018.
"Textile Industries Face Difficulties and Request Urgent Recovery" [electronic reference], on the Syrian Economic News website, at https://goo.gl/fCTPy7, published on December 11, 2016, accessed on July 5, 2018.

from previous generations in most cases, and the long time needed to master very difficult and intricate crafts. The number of craftsmen working in the field of traditional Damascene handicrafts registered in the Union of Craft Associations in Damascus has decreased from 250 in 2011 to 65 in 2018[69].

All of the above factors threaten a number of these crafts with potential extinction if the situation continues as it is, unless quick and serious steps are taken to prevent it[70]. In this regard, Khaldoun al-Masuti, head of the economic office at the Damascus Union of Craft Associations, said that "handicrafts and traditional crafts are the first sector affected by the war and the last one that would be restored to its former glory after the war is over. The sector is a luxurious craft and needs strong mechanisms for export and a stable financial situation to buy. The loss of workers and craftsmen as a result of migration or choosing technology-based occupations threatens the Syrian economy with destruction—something that requires urgent and sustained support"[71].

In the following paragraphs, we will study the changes affecting the three chosen crafts and the difficulties they have faced during the war—i.e. in the period from 2011 until 2018. The study will be classified into three categories—labour, raw materials and machinery, and the outlet market—in order to know precisely the difficulties, challenges and changes that have affected each one alone, although they are very similar in the selected crafts. For a sampling of specific data on changes that have occurred in the crafts under study, kindly refer to Appendix 2, which includes a presentation of some of the quantitative results we obtained during the research.

---

[69] Zeina Shahla (the author), "The Reality of the Damascene Crafts Before and After the War", an interview with Khaldoun al-Masuti of the Damascus Union of Craft Associations.

[70] "States Steal Syrian Traditional Crafts and Put Them into Their Heritage" [electronic reference], on the Sputnik website, at https://goo.gl/obPkaK, published on February 27, 2018, seen on July 5, 2018.

[71] Zeina Shahla (the author), "The Reality of the Damascene Crafts Before and After the War", an interview with Khaldoun al-Masuti, of the Damascus Union of Craft Associations.

1. Changes in Crafts in Terms of the Labour Force

Although the consequences of the war in Syria have affected all aspects of the economy, including all traditional handicrafts, our research found that handicrafts were the sector most affected and most endangered in terms of their continuity and their transmission from one generation to another. Syria's human resources have experienced an unprecedented bleeding as a result of dislocation, death and injuries, which poses a great challenge to the workers in these trades and those who want to recruit young people to teach them the secrets of these professions. An article published in *Al-Ayyam*, a newspaper in Damascus, reports that Syria has lost about 80 percent of its handicrafts professionals—an estimated figure because of the lack of official statistics on the migration of skilled technical and artistic professionals[72]. In this regard, one of the traders selling traditional crafts in al-Khayyateen market said: "There is no new production now that we are selling out of stock, but what will we sell five years from now if it continues as it is? The biggest disaster we have is young people's taking up arms or going abroad"[73]. The same idea is confirmed by Abdul Kareem al-Aseel, who owns a brocade factory and a sales shop in Bab Sharqi. He pointed out in an interview that his sales were only of items from stock, without any new production[74].

In the following paragraphs we will review the changes seen in the three selected crafts in terms of the labour force.

A) Changes in Brocade Production in Terms of the Labour Force

This craft requires a lot of patience and mastery, and its financial return is not rewarding since its production is slow. The return does not currently exceed 5,000 Syrian pounds ($11) per day. Therefore, only those

---

[72] "The Heaviest Loss of War: Skilled Craftsmen and Professional Workers Fled the Country" [electronic reference], on Hashtag Syria website, at https://bit.ly/2GA1aPn, published on October 9, 2017, accessed on July 4, 2018.

[73] Zeina Shahla (author), "The Reality of the Damascene Textile Crafts", an interview with Samer al-Nokta.

[74] Zeina Shahla (author), "The Reality of Brocade Craftsmanship", an interview with Abdul-Kareem al-Aseel, Damascus, [n.p.], on May 9, 2018.

who really love this craft and believe in the importance of preserving it as part of Damascene identity keep working in it[75].

As mentioned in the first part of this paper, the production of brocade goes through several stages, most of which require great precision. Today, this need for precision has become one of the most important working conditions, taking into consideration the high price of silk, which now exceeds $100 per kilogram, compared to $70 before 2011. The high cost of any possible mistake during work, along with the lower sales, means a decline in profits. The number of those wishing to learn the stages of brocade production was in decline even before the war, but the war and its consequences—including poor living and economic conditions and the relative lack of security in Damascus and its surrounding area—contributed to an increasing reluctance among young people to learn this craft. In general, the number of workers in some of the industry's stages today can be counted on fingers[76]. Here, we will mention the workers within several sensitive stages:

1) Dyeing: With the high prices of imported silk yarns, any mistake in this process would be very costly, so many people refrain from working at this stage. In fact, there is only one person, named Mohammed Rehawi, who is known to be practising this profession in Damascus today. He works on the rooftop of a building in the Old City after he lost his workshop in the town of Ein Tarma in the Eastern Ghouta region. This stage of the brocade industry requires primarily a wide area for the purpose of boiling the silk in water and certain materials to remove a sticky gum from the fibres[77] and the preparation of certain dyes. Despite the difficulty of this work and the delicacy it requires, where any mistake would be very costly and

---

[75] Zeina Shahla (author), "The Reality of the Damascene Textile Crafts", an interview with Samer al-Nokta.

[76] This information was obtained through field monitoring of workers in all stages of the brocade industry and interviews with those who are still in Damascus.

[77] According to the interview with the silk dyer Mohammed Rehawi, the price for a kilogram of silk without gum is $200 and the price for silk with gum is $100, so traders tend to buy silk with gum and remove it themselves during the dyeing stage.

might lead to the loss of large amounts of silk[78], still this dyer is not afraid of the craft going extinct. "I learned the profession from my father and I am teaching it today to my 15-year-old son, and there must be people who would like to learn it later," Rehawi said. "Our profession is tough, requires patience, and has a reasonable return. Although I have taught a number of young people, they have abandoned it. The youth of this generation do not have the patience to learn such difficult trades. Nevertheless, I do not lose hope"[79].

Photo 9: Silk Dyeing Tools - Damascus - Zeina Shahla– 2018

2) Punching the Jacquard cards: It has been reported that one person, named Mohamed Saeed Mardini, is still working as a card-puncher in his workshop near Maktab Anbar in Old Damascus. This craftsman believes that this art is about to go extinct, with the tendency to use the electronic textile looms that do not need cards. "I do not think that manual brocade looms will come back to life, for

---

[78] According to the interview with the silk dyer Mohammad Rihawi, the dyeing of 25 kilograms of silk takes four days and is enough to produce about 15 metres of brocade.

[79] Zeina Shahla (author), "The Reality of Silk Dyeing", an interview with Mohammad Rehawi, Damascus al-Qaimariya, [n.p.], on March 18, 2018.

nothing can restore its past status," he said. "There were three of us who worked on punching the loom cards. The other two died and their sons left the profession, despite their knowledge of it. Modernity has eliminated our profession. Brocade in particular has become a folkloric profession that does not make a profit for its professionals. Perhaps this is one of the reasons I do not intend to teach my children this profession, which I inherited from my father. It is in the extinction phase, because of technology first, and war in the second place"[80].

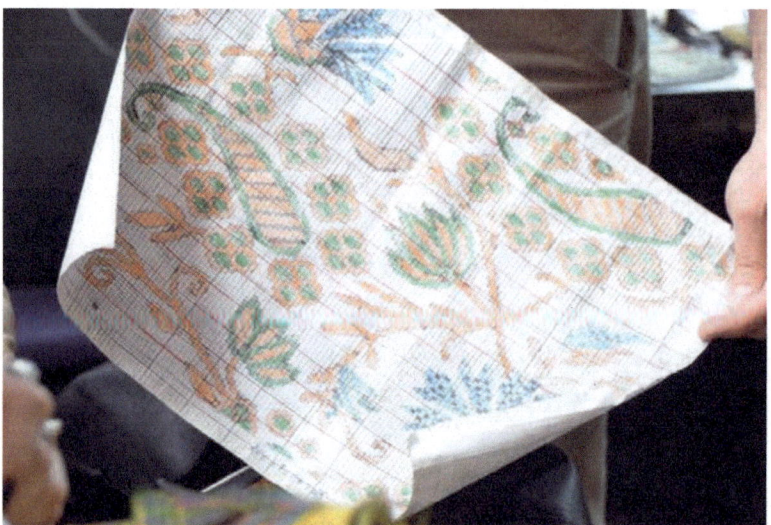

Photo 10: Brocade designs on paper before being transferred to computer - Damascus - Maher Al Mounes – 2018

---

[80] Zeina Shahla (author), "The Reality of Brocade Craftsmanship", an interview with Mohamed Saeed Mardini, Damascus, [n.p.], on May 6, 2018.

Photo 11: Brocade designs on the computer - Damascus - Maher Al Mounes – 2018

3) Weaving: A limited number of people are still working in the weaving of brocade in Damascus, and no new workers have been found. In 2014, the last class of students graduated from the Department of Brocade at the Second Industrial Secondary School in Damascus. The following years have not seen any students in this department, compared to a small number of students in the previous few years, due to the lack of desire to learn a profession that has no future and the lack of any incentives to learn it. Brocade craftsmanship is not attractive to students and their families, which play a role in their sons' selection of their future careers. The deterioration of economic conditions in Syria in general pushed people to look for higher-paying jobs to ensure a better standard of living. Perhaps the most notable evidence of the lack of interest in this craft can be seen in this example: More than 100 students graduated from the Brocade department in this second-

ary school since the department was established in 1994, yet most of them went on to practise other professions[81].

Photo 12: Brocade looms out of work in the Second Industrial Secondary School - Damascus - Zeina Shahla - 2018

Mohammed Rankoussi is one of those who works in brocade weaving and production and is a graduate of the Second Industrial Secondary School and the Middle Institute of Textile Industries. In an interview, he talked about the difficulty of teaching this profession's various stages and the scarcity of those who want to learn. "I can teach brocade weaving, a stage I have mastered after studying it for years," he said, "but I cannot find anyone who wants to learn this craft. Most young people are busy today with quicker-to-learn technology and professions. Besides, there is a lack of awareness of the importance of preserving our heritage from extinction, for which learning this heritage may be the first step to achieve it. We, the Damascenes today, do not know as much about the secrets of our Damascene crafts as we should, and this is where the role of raising awareness [comes in], and creating the desire among young people to learn the secrets of the brocade

---

[81] Zeina Shahla (author), "The Reality of Brocade Craftsmanship", an interview with the Workers at the Second Industrial Secondary School's Textile Department, Damascus, [n.p.], on April 5, 2018.

craft, the stories of its many designs, and thus to feel a sense of belonging to it and contributing in its protection"[82].

Ibrahim al-Ayoubi, a producer of brocade in Damascus, shared the same opinion. In an interview, al-Ayoubi talked about the importance of teaching traditional Syrian crafts to young people, starting in the elementary schools. "If the youths are going abroad or are not willing to learn, then the only hope lies in children to protect our crafts from dying," he said[83].

"The craft is at the head of the textile industry in the world," said another of our interview subjects, Abdul-Kareem al-Aseel, who owns a brocade factory and a sales shop in Bab Sharqi. "It is highly intricate and aesthetic, and it requires a lot of time to master it. A brocade worker needs almost a year and a half to start producing in a professional manner. A mistake in brocade is very costly," he added. "If we would have no choice, we would rely on those who work on the looms that produce the silk textile[84] or the bath cloth 'bashkir'[85], as they have gone a long way in the work on the looms and need to acquire some additional skills. I think the main problem is the craftsmen. We have lost so many of them because of the war"[86]. This comment confirms the same idea regarding the heavy loss suffered by the brocade industry at the level of the labour force during the war, and the need to tackle this problem before the industry loses its last practitioners.

Antoun Mouzannar, one of the most important producers of brocade in Damascus, also talked about the loss of the labour force, but he considered this a general problem in Syria and not only in the manufacture of brocade. He was hopeful of the ability of this industry to regain its lustre once the war is over. "We cannot judge the reality of an

---

[82] Zeina Shahla (author), "The Reality of Brocade Craftsmanship", an interview with Mohammed al-Rankoussi.

[83] Zeina Shahla (author), "The Reality of Brocade Craftsmanship", an interview with Ibrahim al-Ayoubi, Damascus, Sarouja, [n.p.], on March 19, 2018.

[84] The silk textile is a square-shaped embroidered cloth made of cotton and silk. It is used in folk costume especially in the Levant and is worn as a headdress.

[85] The bath cloth "bashkir" is a projecting cotton cloth woven in a special way that allows it to absorb water and moisture and is used to dry the face, hands and body. Syria is well known for its manufacturing of this cloth.

[86] Zeina Shahla (author), "The Reality of Brocade Craftsmanship", an interview with Abdul-Kareem al-Aseel.

industry by studying it during the war. ... Now, there is a problem with the labour force in all fields, but when the war ends, everyone will return," he said. "There is a proverb we usually remember in the market that says, *Whatever dies cannot be revived again, except for the market; for it dies and can live again*"[87].

B) Changes in the Aghabani Production in Terms of the Labour Force

The transformation of Damascus' Eastern Ghouta region into a conflict zone since 2012 has led to a huge change in the reality of the Aghabani craftsmanship. The area has been closed and it became impossible to reach the female workers living there, while hundreds of machines and goods remained trapped inside the region. A number of female workers moved with their machines to live in Damascus and its surroundings and continued to work in embroidery.

Today, it is difficult to count the number of women working in the embroidery of the Aghabani fabric in Damascus. Most of them work at home, while very few work in workshops in the old markets of Damascus. Efforts are under way to educate new female workers in this craft because of its relative ease and the possibility of achieving a reasonable return.

"I learned the profession when I was young. In recent years, I have taught some women in cooperation with the Women's Union in Damascus," said Um Mohammed, a woman who began working in al-Aghabani some 40 years ago. "Many women want to learn, as a good Aghabani set can be produced within a week with a financial profit of up to 10,000 Syrian pounds ($22) per set, thus achieving some kind of self-sufficiency and ensuring the cost of living, especially with the lack of the breadwinner in many cases"[88]. So, the craft of Aghabani embroidery does not seem to be threatened with disappearance because of the presence of dozens of women who master it and can be taught without much difficulty.

---

[87] Zeina Shahla (author), "The Reality of Brocade Craftsmanship", an interview with Antoun Mouzannar, Damascus, [n.p.], on March 18, 2018.

[88] Zina Shahla (author), "The Reality of the Aghabani Craftsmanship", an interview with Um Mohammed, an Aghabani worker, Damascus, [n.p.], on April 28, 2018.

Similarly, a number of girls have been taught at the Women's Union centres or women's associations and also through individual efforts in different areas of Damascus after 2011. Ahmed al-Sheikh, one of the supervisors of such courses, said it was the first time for girls from Damascus to learn the embroidery of the Aghabani, and that this is perhaps one good thing to come from the war and displacement, which contributed to widening the spread of this craft and preventing its extinction[89].

As for the printing process, during the fieldwork we were able to count only two people in Damascus who have mastered it and practise it on almost a daily basis, along with some of the Aghabani producers who can work on it. According to one of the Aghabani producers, who owns a sales shop in al-Khayyateen market: "The number of people working in printing today is sufficient to meet the current demand. If the demand increases, we will see the return of some of the printing professionals or new people will learn the profession. When the demand decreases, they will go to other professions"[90].

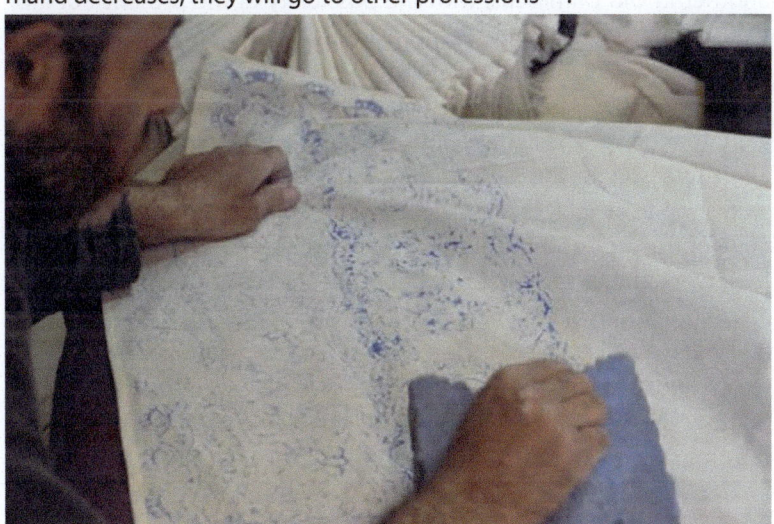

Photo 13: Aghabani's Printing Professional - Damascus - Maher Al Mounes – 2018

---

[89] Zeina Shahla (author), "The Reality of the Aghabani Craftsmanship", an interview with Ahmed al-Sheikh, Damascus, [n.p.], on April 28, 2018.
[90] Zeina Shahla (author), "The Reality of the Aghabani Craftsmanship", an interview with Safouh al-Mawli.

According to Muhiuddin al-Aswad, an Aghabani printing professional who works in a shop in Nazlat al-Shama'een's Medhat Pasha District, there were about eight printing professionals in Damascus and its countryside before 2011, but most of them have left. At the same time, there is no huge production that calls for many workers at the moment. "Before the war, I used to work for about six to seven hours a day," he said. "Now, it has been reduced to three hours per day at most. It's a difficult and an easy job at the same time, and I can teach it to others in a month, but nobody wants to learn it, for it gives no rewarding financial return"[91].

C) Changes in Handmade Rugs Production in Terms of the Labour Force

As a result of the war, the number of looms for producing handmade rugs in Damascus and its surroundings decreased. We were able to count a single workshop in al-Takiyya al-Sulaymaniyah District and another in Old Damascus, with the possibility of the presence of individual looms in some houses but without clear production in the markets[92]. The alternatives for handmade rugs are the machinery produced rugs that do not enjoy the same aesthetic appeal, according to a number of workers in this field.

The reason for this, according to Ghassan Warde, the owner of the workshop at al-Takiyya al-Sulaymaniyah, can be attributed to the difficulty of the profession. "This profession breaks one's back and requires long working hours of up to 10 hours every day, with very little material profit that can hardly meet the daily needs," he said. "For example, in a day and a half I can produce a carpet that costs me around 4,000 Syrian pounds ($9) before I sell it for 6,000 Syrian pounds ($13). Here is another dilemma: A worker who does not own a shop will have to sell it at a lower price to the shops, and therefore the return will not be

---

[91] Zeina Shahla (author), "The Reality of the Aghabani Craftsmanship", an interview with Muhiuddin al-Aswad, Damascus, [n.p.], on May 1, 2018.
[92] Zeina Shahla (author), "The Reality of Handmade Rugs Handicrafts", an interview with Ghassan Warde, owner of a workshop producing handmade rugs, Damascus, on February 24, 2018.

rewarding at all. As for me, the return is fairly enough as I work and sell in my own shop"[93].

The migration of craftsmen to other countries is another reason for concern mentioned by a worker in one of the shops selling rugs in Bab Sharqi. "Today, this and other Damascene handicrafts are threatened with extinction because of migration," he said. "Most of the major manufacturers left and preferred to pursue their profession outside the country, which is the main reason for any possible collapse of this craft"[94].

Despair might also cause today's craftsmen to refrain from teaching the secrets of their crafts to new practitioners, according to a vendor of traditional crafts souvenirs in Bab Sharqi. "Some may want to teach their profession to new generations, but what is the goal? There is no sale and therefore there will be no practice. Teaching this craft will be of little use," he said[95].

2. Changes in Crafts in Terms of Raw Materials and Machinery

The war's drawbacks also affected the machines and tools used in the three selected crafts. Some of these machines were destroyed or stolen, with some factories and workshops located in conflict zones, especially in the Eastern Ghouta region, while others stopped completely due to the lack of production. The lack of craftsmen has also affected the possibility of producing new tools, such as the wooden Aghabani templates.

As for the raw materials, the local production of fabrics was affected in particular by the destruction of some of the largest textile and cloth factories in Damascus and its countryside, as well as the increased prices of imported materials and the difficulty of finding some of them in markets due to the difficulty of importing them. Conse-

---

[93] Zeina Shahla (author), "The Reality of Handmade Rugs Handicrafts", an interview with Ghassan Warde.
[94] Zeina Shahla (author), "The Reality of the Damascene Textile Crafts", an interview with an owner of a shop selling handmade rugs, Damascus, [n.p.], on April 18, 2018.
[95] Zeina Shahla (author), "The Reality of the Damascene Textile Crafts", an interview with an owner of a traditional crafts souvenir shop, Damascus, [n.p.], on April 18, 2018.

quently, the price of the final product has increased, thus reducing the possibility of people buying it, especially local customers.

In the following paragraphs, we will review the changes in the three selected crafts in terms of the raw materials and machinery.

A) Changes in Brocade Production in Terms of Raw Materials and Machinery

After the closure of Syrian silk factories before the war, brocade producers tended to rely on silk and synthetic threads imported from China, mainly without any struggle to obtain them. But, with the increase in silk prices, the cost of production increases significantly. The price for a metre of seven-colour brocade hits about 30,000 Syrian pounds ($65), an increase of more than tenfold, taking into account the low exchange rate of the Syrian pound against the US dollar during the war[96].

Regarding the looms, there are no more than 10 manual brocade looms within Damascus, which are located in several places, including al-Takiyyah al-Sulaymaniyah, al-Sarouja and the Second Industrial Secondary School, in addition to some looms in non-functioning factories such as Mteini factory in Jaramana. Other manual looms were destroyed as a result of their presence in active combat areas such as Joubar, Qaboun and other districts. As mentioned in Part I of this paper, brocade producers have tended to use the automated power looms for many years. During the war, the work of manual and some power looms also stopped, and a few of them are run only upon need. "I still have several looms in my factory, but why should I operate them all?" said Antoun Mouzannar, the brocade factory owner. "I operate a single loom by three workers—out of nine before the war—to produce what I can sell and wait for the war to end," he added. "However, we refuse to stop working permanently and believe in the importance of keeping up, perseverance and continuity"[97].

---

[96] For this information, we have relied on several interviews with brocade shop owners and Damascene textile shops, all of which are among the research's documented interviews.

[97] Zeina Shahla (author), "The Reality of Brocade Craftsmanship", an interview with Antoun Mouzannar.

In any case, there is no difficulty in obtaining new brocade looms upon need, but that need is absent today with the scarcity of production.

B) Changes in the Aghabani Production in Terms of Raw Materials and Machinery

Production of cloth at al-Khomassiya and al-Maghazil factories, mentioned in Part I of this paper, stopped because of their location in conflict areas. This led to a decline in the availability of these fabrics for use in al-Aghabani sets and the need to rely on alternatives that are not of the same quality and aesthetic appeal, or, preferably, to keep relying on old stock if available, according to the workers in Aghabani production we interviewed in this research.

The threads used in producing the Aghabani embroidery are divided into two main types: silk threads, which are imported from Poland or China now instead of France, and synthetic threads, whose main source is China or India. Aghabani producers do not worry much about finding the threads because of the current lack of demand for production. However, the prices have risen from $5 per kilogram before 2011 to $8 per kilogram this year. "Threads are available because there is not much production to consume so much," one producer said. "The threads available in the market are commensurate with the amount of what we produce"[98].

After 2011, most of the Aghabani machines remained in Eastern Ghouta, which had survived harsh battles over the years of fighting, but some of the women practising the Aghabani craft managed to take their machines out of the region and continue working elsewhere. It is difficult to know the fate of the machines that are still in Ghouta, where some of them are likely to have been destroyed or stolen.

The machines available in the market today are mostly used ones priced between $300 and $500, and new machines of Taiwanese or Chinese origin costing about $1,000. People who wish to work on Aghabani machines can purchase or rent them, taking into consideration that the old machines are preferred by experts in terms of quality

---

[98] Zeina Shahla (author), "The Reality of the Aghabani Craftsmanship", an interview with Safouh al-Mawli.

over the new ones. It is estimated that 1,000 machines are still in use almost daily in Damascus[99].

All of the wooden printing moulds available today are used ones, with no new moulds being produced. This was confirmed by all the Aghabani workers we interviewed, and through a wide tour of the workshops for the manufacture of wooden crafts and the shops selling various wooden moulds in Damascus' old neighbourhoods and the Qabqabiya market.

With some spare moulds carrying each of the Aghabani's designs, it is possible to carry on like this for up to six years before a new mould industry is needed. "If I have to, I will try to make it myself," said Muhiuddin al Aswad, the Aghabani printing professional. "I have no alternative anymore"[100].

C) Changes in the Production of Handmade Rugs in Terms of Raw Materials and Machinery

The handmade rugs industry was not very affected in terms of the raw materials and machines. This is because of its reliance on cotton and woollen yarns, obtained from local sources and mostly from Aleppo, or from an external source like China. There are no major problems in acquiring them today. The machines used are wooden looms and their accessories, and it is easy to design a similar loom upon need[101].

3. Changes in Crafts in Terms of the Outlet Market

The negative impacts we have seen in the previous paragraphs on the labour force and on raw materials and machinery have also taken a toll on the outlet market for the selected crafts. The total sales of all shops selling traditional handicrafts and souvenirs decreased sharply due to the high prices of their products and the lack of tourists, who are the principal customers for these products.

---

[99] Zeina Shahla (author), "The Reality of the Aghabani Craftsmanship", an interview with Ahmed al-Sheikh, Damascus.
[100] Zeina Shahla (author), "The Reality of the Aghabani Craftsmanship", an interview with Muhiuddin al-Aswad.
[101] Zeina Shahla (author), "The Reality of Handmade Rugs Handicrafts", an interview with Ghassan Warde.

In the following paragraphs, we will review the changes in the three selected crafts in terms of the outlet market.

A) Changes in the Brocade Industry in Terms of the Outlet Market[102]

Today, the market for brocade products has changed dramatically, with sales dropping by 70 percent at the very least since before the war.

Currently, this market depends mainly on customers residing outside of Syria, or on sales that are generally directed abroad. A number of brocade producers estimate their product distribution rate as 15 percent for high-income people internally, and 85 percent for the outside market. The proportions differ slightly according to each of the shops we visited. The owners of the shops link the low percentage of sales inside the country first of all to the security situation, which is one of the most important conditions for the prosperity of handicrafts, as well as to economic considerations, as local customers look for less expensive gifts and products, and brocade is not likely to be among them. Moreover, the lack of the necessary marketing and awareness of the importance of this craft and its association with Damascene identity also affected the industry. These sellers also consider that the decline or even the lack of pre-eminence in the demand for brocade products, which previously were sold almost on a daily basis and mostly abroad, is an important indicator of the decline in sales. This prompted some vendors, who previously specialised in selling brocade products, to display other fabric products in order to continue working.

In this regard, Abdul Kareem al-Aseel compares his sales before and after the war. "Before the war, I had a permanent stock of about 8,000 metres of brocade, a figure that was constant over the years," he said. "As production halted after the war, I have only managed to sell 2,000 metres of this stock over the past years without producing anything

---

[102] For this information, we have relied on interviews with a number of brocade producers and vendors in Damascus.

new. One of the main reasons for this is the weakness of export to external markets"[103].

B) Changes in the Aghabani Production in Terms of the Outlet Market

There are at most about ten shops that specialise in the sale of the Aghabani in Damascus. They are the same shops that were operating before the war, and whose owners insist on fighting conditions and continuing to work, including the shops of al-Mawli, Sukkar and al-Rifa'i in al-Khayyateen market. In general, owners of crafts souvenir shops in the traditional markets are keen to display the various products of Aghabani, which is considered one of the most desirable Damascene textile products, as customers from various social and economic classes inside and outside Syria want to acquire it, recognising that it is closely related to the Damascene identity. In addition, its price is relatively reasonable. There is hardly any Damascene house, or any house of a Damascene person living in the Diaspora, that does not have at least one piece of Aghabani. "Aghabani is always a desirable product and I cannot imagine my house without the Aghabani tablecloths, especially the ones we use at social events and when guests arrive," said one of the Aghabani producers and vendors in al-Khayyateen market[104].

As a result of the availability of all production elements, namely the labour, raw materials and machinery, the volume of Aghabani production depends heavily on sales. "In the past, we were able to complete 100 sets in two weeks or 1,000 sets in two weeks, where I could expand into the region and find new female workers in the market upon need," said an Aghabani producer. "Today the equation is still the same. Production and sales decreased proportionally by more than 80 percent. We barely produce a few sets, no more than ten per week, which is sufficient to meet the current sales traffic, which mostly targets people, wholesalers or fairs outside of Syria and much less widely

---

[103] Zeina Shahla (author), "The Reality of Brocade Craftsmanship", an interview with Abdul-Kareem al-Aseel.

[104] Zeina Shahla (author), "The Reality of the Aghabani Craftsmanship", an interview with Safouh al-Mawli.

for people living inside the country. Otherwise we would face a major depression in goods. We can sell what we produce now"[105].

C) Changes in the Handmade Rugs Industry in Terms of the Outlet Market

With the decline in the production of handmade rugs in Damascus, few shops continue to sell these products. In a survey we conducted, only three shops selling handmade rugs were found in al-Takiyya al-Sulaymaniyah, Bab Sharqi and Medhat Pasha. Most of the shops selling rugs—there are dozens of them—tend to sell rugs made using automatic looms, which are woven with more precision but have less aesthetic appeal.

According to the owners of these shops, sales decreased by more than 90 percent after 2011, and this can be attributed mainly to the lack of tourists and the difficulty of exporting, especially to Jordan and Turkey with the closure of those countries' borders with Syria. Today, vendors rely on foreign markets for their sales. "For years, we spent most of our days in the shop without seeing a customer," said a vendor in Bab Sharqi. "However, I insist on opening my shop every day just to avoid the rumour that I have closed it. I wait for stability, for it is our only hope for returning our work to how it was before the war"[106]. One of the reasons for the decline in the sales of handmade rugs, according to the owner of the Qashlan shop in Medhat Pasha, is the rise of the price of handmade rugs to twice the price of the machinery-made rugs, making the latter more desirable, especially for local customers[107].

For Ghassan Warde, the handmade rugs workshop owner, the reason for the decline in the sales of handmade rugs is the decline in "the purchasing ability of Syrian citizens that led workers to abandon these crafts that are no longer materially viable"[108].

---

[105] Zeina Shahla (author), "The Reality of the Aghabani Craftsmanship", an interview with Safouh al-Mawli.
[106] Zeina Shahla (author), "The Reality of the Damascene Textile Crafts", an interview with an owner of a traditional crafts souvenir shop.
[107] Zeina Shahla (author), "The Reality of Handmade Rugs Craftsmanship", an interview with the owner of Qashlan Handmade Rugs Store, Damascus, [n.p.], on May 9, 2018.
[108] Zeina Shahla (author), "The Reality of Handmade Rugs Handicrafts", an interview with Ghassan Warde.

It is interesting to note that a number of other Damascene textile selling shops have also been selling machinery-made rugs in recent years. They are relatively low-priced goods, and can be a reasonable source of income, compared to other handmade Damascene products whose costs and selling prices have increased significantly.

4. Local and Regional Projects and Initiatives Launched to Confront the Risk of Extinction of Traditional Crafts

As seen in the previous paragraphs, the Damascene handicrafts have been affected by the country's ongoing war since 2011. This was evident from the survey we conducted to complete this research, and through interviews that showed the impact on most of these trades and the urgent need to protect them from the risk of extinction. That risk stems mainly from a decline in the expert labour force, the destruction of workshops and factories, and the loss of places necessary for the continuation of the work of artisans.

Ibrahim al-Ayoubi compared these handicrafts to the Syrians' "passport" by which they assert their identity to the world. However, he said he believed that new craftsmen were needed and that governmental bodies needed to take responsible and practical measures to protect the Syrian crafts from extinction[109].

The work to preserve and restore these crafts requires concerted formal and informal efforts to support them and encourage their workers to continue practising and developing their skills[110]. In this regard, a study titled "The Brocade Industry and the Significance of Brocade-Related Decorations" discusses the difficulties facing the brocade industry. The weakness of marketing is a direct cause of this in-

---

[109] Zeina Shahla (author), "The Reality of Brocade Craftsmanship", an interview with Ibrahim al-Ayoubi, Damascus.

[110] See the following articles on the need to protect Syria's handicrafts from extinction and loss:
Hisham Adra, "Syria's Handicrafts and Traditional Knowledge at Risk of Loss" [electronic reference], on *Asharq al-Awsat* website, found at https://goo.gl/28eA1y, published on February 3, 2015, accessed on December 10, 2019.
Bassam al-Mustafa, "al-Qalaa Warns of the Disappearance of Folklore Crafts and Appeals to Artisans to Return Home" [electronic reference], on World and Syria news website, at https://goo.gl/4CfDy5, published on December 29, 2016, accessed on July 5, 2018.

dustry's decline and an impediment to achieving its sustainability, the study concluded. Its results also showed the extent of threats to the profession, especially the emigration of artisans, and warned decision-makers to resist those threats and find solutions to help the brocade industry rise again[111].

Realising the large losses that have gradually affected these crafts, many formal and informal initiatives have emerged at the local and regional levels, with the aim of preserving and protecting such crafts by supporting education, production and marketing through various ways and means. Since 2011, many associations and projects have been established, each of which has taken a specific path to achieve these goals. Some of them have sought to teach handicrafts to new workers. Other efforts have tried to provide material and logistical support to crafts practitioners, especially to those who lost their workplaces or suffered from significant material losses. Other initiatives documented aspects of Syria's intangible cultural heritage to protect it from loss, as well as raise awareness of the importance of protecting this heritage, which forms an essential part of Syria's identity.

Examples of such initiatives include the handwork incubator at the Glass Factory in Dummar District, Damascus, which was established in 2017 with the aim of developing traditional crafts under the supervision of the General Union of Craftsmen and the Ministries of Tourism, Culture and Economy; the "Khuyut al-Amel" project established in Damascus in 2012 to preserve the Aghabani crafts by producing new Aghabani pieces with modern and contemporary designs; the Sama project, which was founded in Damascus in 2014 and seeks to support and develop the traditional Syrian crafts; the Abha project, which was established in 2013 with the aim of preserving the non-material Syrian cultural heritage represented by the Syrian handicrafts from extinction; and "Sham el-Yasamin," an association for Damascene cultural heritage that was established in 2011 to preserve old crafts from extinction.

In addition, there are a number of programmes and projects that are working in cooperation with UNESCO to preserve the traditional

---

[111] Salem al-Ahmad Abdullah, "The Brocade Fabric Industry and the Significance of Brocade-Related Decorations", a master's thesis in Rehabilitation and Specialization in Folklore, Damascus University, Faculty of Arts and Humanities, Department of Sociology, Damascus, 2015/2016, p. "K".

Syrian crafts, including the National Program for Sustainable Development of Handicrafts, which aims to revive some endangered traditional crafts by training a new generation to develop and innovate new methods, and the International Observatory of Syrian Cultural Heritage, which monitors damage and threats to the nation's built, movable and intangible cultural heritage, combats the illegal trafficking of stolen or looted objects, and collects information that will help in restoring the nation's cultural heritage[112].

In parallel, the Syrian Ministry of Tourism established the Higher Council of Traditional Crafts in 2013, which aims to revive and preserve traditional professions and industries in cooperation with the relevant public and private bodies by providing the necessary technical, administrative and financial requirements to activate this sector[113].

However, most of the Damascene textile workers we interviewed believe that these efforts are not enough and that they need more support so as not to remain mere ink on paper, and in order to exercise their role to take these crafts to a safe shore. This requires the appropriate activation of these initiatives to meet the current situation in the country and the reality of artisans' work and actual needs.

For example, Khaldoun al-Masuti, the head of the Union of Craft Associations' economic office in Damascus, talked about experiments conducted in cooperation with UNESCO and the Ministry of Tourism to teach some handicrafts to a number of young people who wish to learn them. However, he said that these experiments are not enough and require long practice to achieve the desired impact[114].

For Mohammed Rankoussi, the brocade professional, this craft sounded the alarm, with no formal or informal body seriously helping

---

[112] UNESCO, "The Emergency Safeguarding of the Syrian Cultural Heritage Project" [electronic reference], on UNESCO's website, at http://bit.ly/2yMuRYl, (n.d.), accessed on August 6, 2019.

[113] Ameer Sabor, "The Higher Council of Traditional Crafts' First Meeting Emphasises the Importance of Preserving Traditional Professions and Crafts" [electronic reference], on Syria Steps website, at http://www.syriasteps.com/?d=134&id=106587, published on June 29, 2013, accessed on January 10, 2019.

[114] Zeina Shahla (the author), "The Reality of the Damascene Crafts Before and After the War", an interview with Khaldoun al-Masuti.

to preserve it[115]. Similarly, Ghassan Warde, the owner of the workshop at al-Takiyya al-Sulaymaniyah, said that workers making handmade rugs do not find any appreciation for their efforts either from official bodies or from people who have lost interest in handicrafts for many reasons[116].

All of these findings highlight the importance of doubling the exerted efforts and most importantly organising them in a way that ensures the achievement of the desired goal of protecting all aspects of the Syrian cultural heritage.

---

[115] Zeina Shahla (author), "The Reality of Brocade Craftsmanship", an interview with Mohammed al-Rankoussi, Damascus.

[116] Zeina Shahla (author), "The Reality of Handmade Rugs Handicrafts", an interview with Ghassan Warde.

Conclusion and Recommendations

Each Syrian artisanal handicraft has its own specific identity stemming from its history, location, and the stages of work and the elements involved in its production. Together, however, these crafts constitute a very important part of the Syrian cultural heritage, which should be preserved in times of peace and war. This research was a journey to explore some features of the current reality of a limited number of Damascene artisanal crafts and how they were affected at various levels by the war that has raged in Syria since 2011.

We sought to achieve this through selecting three models of such crafts, namely brocade, the Aghabani and handmade rugs, and discussing the details of the changes that have occurred to each of them in terms of labour force, machinery, raw materials and sales markets. The results we found speak of a deterioration in the reality of these crafts, and of all Syrian Damascene crafts in general, and especially in terms of the labour force, which witnessed a draining that could be described as catastrophic during the war, through the loss of workers due to death, injury and leaving the country on the one hand, and the preference of many artisans to work in easier and more profitable fields on the other. The deterioration of the Syrian crafts also includes the destruction of workshops and factories, the loss of a large part of the manufacturing machinery, the loss of local sources of raw materials and the difficulty of importing them from abroad, in addition to the almost complete cessation of sales movement in local markets due to the difficulty of exporting and the lack of interest among local customers in acquiring the products of those trades, which are considered luxuries rather than necessities.

While it is difficult to explore the reality of entire Syrian crafts within the temporal and spatial limits of this research, we see that the selected examples represent a sample whose features can be generalised to different crafts, bearing in mind the specificity of each craft, as the suffering in terms of labour or outlet markets is similar. In addition, the steps that must urgently be taken to protect this important part of our heritage from loss are the same for all crafts, and must cover all of them without exception.

Based on the interviews conducted within the scope of this research, a number of recommendations were made by the interviewees. We will mention the following, ranked by priority and classified by the party we believe should be directed to it, taking into account the difference of these recommendations from one person to another, as a result of different working realities and outlooks for the current and future status of the craft they work in.

Recommendations addressed to Syrian governmental entities regarding the support of current crafts and craftsmen:

- Help the artisans to overcome their difficulties and revive their crafts. This aid can take several forms, including providing loans, opening small workshops for those who have lost their workshops during the war or to those who do not have the financial capacity to open their own workshops. One of the craftsmen suggested the idea of repurposing closed factories by using them as a hub for craftsmen, stressing the need to provide these places free or at a nominal price to craftsmen to help and encourage them to continue.
- Pay more attention and appreciation to the craftsmen and workers at various levels and stages of the handicraft professions through several measures, including raising the quality of social services provided to them such as health insurance and care, and providing professional training to raise their skills in various fields, such as business management.
- Secure sufficient salaries for workers in these crafts to keep them working—an idea not agreed upon by all interviewees, where some believe that this might stimulate fraud and dependence.
- Secure the raw materials through official channels at reduced customs duties.
- Besides supporting heavy industries, pay more attention to traditional handicrafts and draw up special plans to preserve them.

Recommendations addressed to Syrian government entities regarding the training and qualification of new labour force members to master handicrafts:

- Establish advanced vocational training centres, whose curricula and contents are directly related to the needs of the labour market, and allocate salaries to trainees during the instructional period to encourage them to join and adhere to these centres.
- Cooperate with higher education institutions and create departments in related faculties, such as the Faculty of Fine Arts, to develop and preserve traditional crafts.
- Encourage young people to learn handicrafts—for example, by establishing an art school separate from the regulations of other schools and institutes where any person who wishes to join such a school is entitled to enrol without conditions related to age or secondary school certificates and grades.
- Encourage the non-governmental sector to support artisans, as this sector can carry out a number of projects and programmes within this context as part of its social responsibility. Such encouragement can take several forms, such as the provision of legal facilities and the reduction of taxes on such projects and programmes.
- Educate women in different stages of the textile crafts to overcome the shortage of men because of the war[117]. Transfer the crafts professions to new generations and divide these generations into two categories: a class that will work for obtaining a financial return and a class that will work for the purpose of art appreciation.
- Do not limit efforts to educate a number of young professionals to training only, but follow up on their progress after the completion of the training courses.

---

[117] Shubat, Abdul-Hadi, "al-Nouri: Women Make Up 60 Percent of the Population and We Need to Rehabilitate Their Leadership Skills", on *Al-Watan* newspaper website, at http://alwatan.sy/archives/88974, published on January 26, 2017, accessed in April 2019.

Recommendations addressed to Syrian governmental entities related to the marketing and support of Syrian crafts products:

- Promote and market craft products inside and outside the country, which is one of the most important keys to supporting traditional crafts, through a set of procedures, including the use of archaeological sites, media and outreach, seminars, exhibitions, commercial markets, participation in foreign fairs at nominal prices, and the establishment of folklore villages and incubators for handicrafts.
- Compel government departments, like embassies, the People's Assembly and others, to purchase traditional local products, to compensate for the decline in the number of tourists who were the main customers for these products.
- Restore the production of silk threads and fabrics locally, thus reducing the cost of production.
- Adopt new industrial strategies, such as directing big industrialists to support handicrafts in different ways, such as buying the products of those craftsmen or allocating part of the taxes to support them.
- Open the doors to export handicraft products through official channels.

Recommendations for Syrian non-governmental entities to support handicrafts:

- Implement programmes and projects aimed at supporting current craftsmen, supporting their continuity to work and assisting in the marketing of their products as part of the social responsibility of these entities.
- Implement programmes and projects aimed at teaching new generations of youths all stages of Syrian handicrafts, as part of the social responsibility of these entities.

Recommendations to Syrian and international entities in terms of the documentation, study and support of handicrafts:

- Pay attention to documentation of various aspects and stages of handicrafts, in order to facilitate efforts to preserve them.
- Raise public awareness regarding the traditional Syrian craftsmanship among all segments of the society, including children, by enriching educational curricula with related materials.

We hope that these recommendations will be useful in the short and long term, and that they will receive the attention of those who are able to implement them from various official and civil bodies and experts inside and outside Syria. We emphasise the need to meet with the craftsmen and workers in these trades at various levels and negotiate with them regarding any new recommendations that can work to support them.

We also hope that this research will open a wider door to ask similar questions about the reality of other Syrian crafts, for each deserves separate research in its own right, or to ask different questions about aspects that we might be unaware of, as well as to propose new solutions for the protection of traditional Syrian crafts—a process that should not stop even after the war is over. The destruction caused by a conflict that has lasted for many years will take decades to rehabilitate and overcome its effects, and we do not wish to reach a time when we lose our crafts and heritage and regret that no efforts to preserve them were made.

References

Interviews

Shahla, Zeina (the author). "The Reality of the Damascene Crafts Before and After the War", an interview with Khaldoun al-Masuti, head of the economic office at the Damascus Union of Craft Associations, [n.p.], Damascus, on June 21, 2018.

Shahla, Zeina (author). "The Reality of Brocade Craftsmanship", an interview with Mohammed al-Rankoussi, Damascus, [n.p.], on January 11, 2018.

Shahla, Zeina (author). "The Reality of the Aghabani Craftsmanship", an interview with Safouh al-Mawli, Damascus, [n.p.], on December 30, 2017.

Shahla, Zeina (author). "The Reality of the Damascene Textile Crafts", an interview with Samer al-Nokta, Damascus, [n.p.], on January 11, 2018.

Shahla, Zeina (author). "The Reality of Brocade Craftsmanship", an interview with Ibrahim al-Ayoubi, Damascus, Sarouja, [n.p.], on March 19, 2018.

Shahla, Zeina (author). "The Reality of the Aghabani Craftsmanship", an interview with Ahmed al-Sheikh, Damascus, [n.p.], on April 28, 2018.

Shahla, Zeina (author). "The Reality of Silk Dyeing", an interview with Mohammad Rehawi, Damascus, al-Qaimariya, [n.p.], on March 18, 2018.

Shahla, Zeina (author). "The Reality of Brocade Craftsmanship", an interview with Mohamed Saeed Mardini, Damascus, [n.p.], on May 6, 2018.

Shahla, Zeina (author). "The Reality of Brocade Craftsmanship", an interview with the workers at the Second Industrial Secondary School's Textile Department, Damascus, [n.p.], on April 5, 2018.

Shahla, Zeina (author). "The Reality of Brocade Craftsmanship", an interview with Abdul-Kareem al-Aseel, Damascus, [n.p.], on May 9, 2018.

Shahla, Zeina (author). "The Reality of Brocade Craftsmanship", an interview with Antoun Mouzannar, Damascus, [n.p.], on March 18, 2018.

Shahla, Zeina (author). "The Reality of the Aghabani Craftsmanship", an interview with Um Mohammed, an Aghabani worker, Damascus, [n.p.], on April 28, 2018.

Shahla, Zeina (author). "The Reality of the Aghabani Craftsmanship", an interview with Muhiuddin al-Aswad, Damascus, [n.p.], on May 1, 2018.

Shahla, Zeina (author). "The Reality of Handmade Rugs Handicrafts", an interview with Ghassan Warde, Damascus, [n.p.] on February 24, 2018.

Shahla, Zeina (author). "The Reality of the Damascene Textile Crafts", an interview with an owner of a shop selling handmade rugs, Damascus, [n.p.], on April 18, 2018

Shahla, Zeina (author). "The Reality of the Damascene Textile Crafts", an interview with an owner of a shop selling traditional crafts souvenirs, Damascus, [n.p.], on April 18, 2018.

Shahla, Zeina (author). "The Reality of Handmade Rugs Craftsmanship", an interview with the owner of Qashlan Handmade Rugs Store, Damascus, [n.p.], on May 9, 2018.

Books

Al-Badri, Abu al-Baqa'. *Nuzhat al-anam fi mahasin al-Sham (A Night Walk in the Beauties of the Levant)*, the Arabic Library, Baghdad, the Salafi Printing House "al-Matba'a al-Salafiyya". Egypt, 1341 (1922). Electronic reference, available at https://goo.gl/aVkDqF.

Zuhdi, Bashir. *Studies in Damascene History, Archaeology and Craftsmanship*. Damascus: Dar Al-Hilal Publishing House and Dar Al-Yanabi' Publishing House, 2010.

Al-Qasimi, Kamal, and Hasan al-Hamami. *A Report on Traditional and Artistic Industries and Handicrafts in Syria*, [n.p.], 1975.

Fayyad al-Fayyad, Mohammed, and Majed Hashim Hammoud. *Traditional Crafts in Syria*, first edition, translated by Majd Hamoud. Damascus: General Union of Craftsmen, Office of Culture and Media, 2011.

Kayyal, Mounir. *Levantine Achievements in Damascene Arts and Industries*. Damascus: Syrian General Authority for Books, Ministry of Culture, 2007.

Theses

Salem al-Ahmad Abdullah. "The Brocade Fabric Industry and the Significance of Brocade-Related Decorations", a master's thesis in Rehabilitation and Specialization in Folklore. Damascus: Damascus University, Faculty of Arts and Humanities, Department of Sociology, 2015/2016.

Marshe Fouad, Nicola. "Study of the Reality of Training Units for the Rugs and Carpets Industry in the Syrian Arab Republic and Evaluation of Their Activities in the Economic and Social Fields", thesis for a Diploma in Economic and Social Planning and Industrial Planning. Damascus: State Planning Commission, Institute of Economic and Social Development Planning, 1982.

Monographs

The Intangible Syrian Cultural Heritage: Skills Associated with Traditional Craftsmanship, Part I. Damascus: Syrian Ministry of Culture and the Syria Trust for Development, 2014.

Documents and Articles

UNESCO. "Convention for the Protection of Cultural Property in the Event of Armed Conflict" [electronic reference], on the UNESCO website, at https://goo.gl/S9HNFy, 2017, accessed on July 5, 2018.

UNESCO. "The Emergency Safeguarding of the Syrian Cultural Heritage Project" [electronic reference], on the UNESCO website, at http://bit.ly/2yMuRYI, (n.d.), accessed on August 6, 2019.

"The General Union of Craftsmen" [electronic reference], a webpage on the economic website of the General Federation of Artisans Associations, at https://goo.gl/wWdbnX, [n.d.], accessed on July 5, 2018.

Onder, Harun, et al. "The Toll of War: The Economic and Social Consequences of the Conflict in Syria" [electronic reference]. Washington, DC: World Bank Group, 2017. Available at http://documents.worldbank.org/curated/en/811541499669386849/full-report.

Al-Aloush, Fadi. "The Ministry of Tourism Announces Support for Handicrafts Through a Specialised Directorate" [electronic reference], on the website of Syria Steps, at https://goo.gl/Sw71Dm, published on November 30, 2011, accessed on July 1, 2018.

"Traditional Industries in Syria: A Confirmation of Our Civilised Identity" [electronic reference], on *Al-Fidaa* newspaper website, at https://goo.gl/d3rVAz, published on December 8, 2011, accessed on January 10, 2019.

"Heritage Products Struggle against Time to Preserve the Identity of Communities" [electronic reference], on the Naba' website, at https://goo.gl/gTD58B, published on October 5, 2010, accessed on July 5, 2018.

Ibrahim, Y. "As Most of Handicrafts Have Become Threatened with Extinction, They Are Attracting Visitors to Tartus Summer Festivals 2010" [electronic reference], on *al-Wehda* newspaper website, at https://goo.gl/Xinqya, published on August 8, 2010, accessed on July 5, 2018.

"Including the Brocade Fabrics, the Aghabani and Ajami, Damascene Handicrafts Are In Danger of Extinction" [electronic reference], on the website of *Al-Ittihad* newspaper, at https://goo.gl/owwAoT, published on January 31, 2011, accessed on July 5, 2018.

Makiyya, Osama. "Abu Mahmoud: The Last Silk Dealer in Damascus and the Profession of Handmade Glass Is Dying" [electronic reference], on Syria News website, found at https://goo.gl/wsHqm7, published on March 26, 2010, accessed on July 5, 2018.

Salman, Hassan. "The Silk Industry in Syria from Prosperity to Decline" [electronic reference], on the website of *Al-Bayan* newspaper, at https://goo.gl/enZbH1, published on October 2, 2009, accessed on July 5, 2018.

"Handicrafts Demonstrate the Skills of Syrian Craftsmen" [electronic reference], on the *Day Press* news website, at https://goo.gl/P9iHJo, published on June 2, 2010, accessed on December 10, 2019.

"Encouraging Traditional Handicrafts Increases Their Contribution to the GDP" [electronic reference], on *al-Jamahir* website, found at http://jamahir.alwehda.gov.sy/node/383894, published on July 10, 2012, accessed on July 1, 2018.

"Syria's Famous Brocade Silk Attracts Tourists despite Rising Price" [electronic reference], Reuters news article, found at https://ara.reuters.com/article/entertainmentnews/idARAOLR2211 5420080722, published on July 23, 2008, accessed on July 1, 2018.

Al-Jammal, Alaa. "Sherko Matini: 'The Brocade', A Damascene Beauty with a Folk Heritage Imprint" [electronic reference], on eSyria website, at https://goo.gl/65KwL3, [n.d.], accessed on July 5, 2018.

Zeitoun, Joseph. "The Damascene Brocade" [electronic reference], on Joseph Zeitoun's website, at: https://goo.gl/KVkPKQ, published on February 26, 2016, accessed on July 5, 2018.

Satik, Sinan. "The Damascene Brocade: The Queen's Cloth Industry Is Endangered" [electronic reference], on the *New Arab* website, at https://goo.gl/sQbKBH, published on August 16, 2015, accessed on July 5, 2018.

Abu Abed, Ammar. "Damascene Brocade Weaves Silk with Silver and Gold Threads" [electronic reference], on the website of *al-Ittihad* newspaper, at https://goo.gl/jBbmsu, published on March 9, 2010, accessed on July 5, 2018.

Abbas, Fatina. "Silk Farming in Tartus: A Long-Standing Career on the Road to Demise" [electronic reference], on *Panorama Tartous* website, at https://goo.gl/djBjM7, published on May 1, 2016, accessed on July 3, 2018.

Haitham, Yahya Mohammed. "The Government Delegation Headed by Engineer Khamis Continues His Field Visit to Tartus Province" [electronic reference], on *al-Thawra* newspaper website, at https://goo.gl/brjwG9, published on April 17, 2017, accessed on July 5, 2018.

Deeb, Ghosun. "Silkworm Breeding in Tartus: The Lack of Marketing and the High Cost of Supplies Are the Most Important Reasons behind the Decline in Production" [electronic reference], on *al-Thawra* newspaper website, at https://goo.gl/qwdKK7, published on January 14, 2012, accessed on July 5, 2018.

Adra, Hisham. "Five Rug-Producing Manual Looms Are Left in Damascus" [electronic reference], on the website of *Asharq al-Awsat* newspaper at https://goo.gl/GmUWJh, published on February 13, 2009, accessed on July 5, 2018.

Adra, Hisham. "Studies Confirm the Feasibility of Reviving Syrian Silk Industry" [electronic reference], on the website of *Asharq al-Awsat* newspaper, at https://goo.gl/FRfoKN, published on April 8, 2001, accessed on July 3, 2018.

Adra, Hisham. "Syria's Handicrafts and Traditional Knowledge at Risk of Loss" [electronic reference], on the website of *Asharq al-Awsat* newspaper, at https://goo.gl/28eA1y, published on February 3, 2015, accessed on December 10, 2019.

"Minister of Tourism: Tourism Losses 387 Billion Syrian Pounds and a Decrease of 98 Percent in the Number of Tourists" [electronic reference], on Syria News, at https://goo.gl/f2QCq5, published on March 23, 2015, accessed on July 5, 2018.

Issa, Sami, and al-Khalil Markazan. "The Reality of the Textile Industry under the Microscope of Experts: The Vision of the Ministry to Provide Financial Support for Rehabilitating the Expert Hand" [electronic reference], on *Tishreen* newspaper website, at https://goo.gl/Sj4bjm, published on July 25, 2016, accessed on July 5, 2018.

"Textile Industries Face Difficulties and Request Urgent Recovery" [electronic reference], on the Syrian Economic News website, at https://goo.gl/fCTPy7, published on December 11, 2016, accessed on July 5, 2018.

"States Steal Traditional Syrian Crafts and Put Them into Their Heritage" [electronic reference], on the Sputnik website, at https://goo.gl/obPkaK, published on February 27, 2018, accessed on July 5, 2018.

"The Heaviest Loss of War: Skilled Craftsmen and Professional Workers Fled the Country" [electronic reference], on Hashtag Syria website, at https://goo.gl/4W24iz, published on October 9, 2017, accessed on July 4, 2018.

Al-Mustafa, Bassam. "al-Qalaa Warns of the Disappearance of Folklore Crafts and Appeals to Artisans to Return Home" [electronic reference], on World and Syria news website, at https://goo.gl/4CfDy5, published on December 29, 2016, accessed on July 5, 2018.

Sabor, Ameer. "The Higher Council of Traditional Crafts' First Meeting Emphasises the Importance of Preserving Traditional Professions and Crafts" [electronic reference], on Syria Steps website, at http://www.syriasteps.com/?d=134&id=106587, published on June 29, 2013, accessed on January 10, 2019.

Shubat, Abdul-Hadi. "al-Nouri: Women Make Up 60 Percent of the Population and We Need to Rehabilitate Their Leadership Skills" [electronic reference], on *Al-Watan* newspaper website, at http://alwatan.sy/archives/88974, published on January 26, 2017, accessed in April 2019.

Appendix 1: Proverbs and Folk Songs Related to the Textile Crafts

The following is a list of proverbs and folk songs related to the Syrian textile crafts:

Proverbs:

1) The thread craft is unbearable (Kar el-taa' ma beyentaa'): El-taa' is the thread in Syrian dialect. The proverb refers to the harshness and difficulty of the weaving profession, and brocade weaving in particular.
2) O you who master the loom work, it is about doing rather than saying (Ye metfannin bel nool, el-shaghle ma bel ool): This proverb also refers to the difficulty of the profession and the fact that it requires work rather than words.
3) Set the beater and the pedal well, and bad luck will fade away (Zabbet el-datte we el-dawse, betroohannak el-manhoose): That is to say, working on the loom (the beater and pedal are parts of it) expels bad luck.
4) Take a beater here and a beater there, and see how [perfect] the work will be: (Khod daffe la hon we daffe la hon, we shof el-sheghelshlon): A saying used to encourage work on the textile loom despite its difficulty.
5) Have you extended the Aghabani [tablecloth]? We have extended the Aghabani for your sake (Sho faradtoo al-Aghabany? Faradni al-Aghabani la oyonkum): A common saying used in Damascus, where homeowners boast of decorating their tables with Aghabani tablecloths to celebrate guests.
6) Warp and weft (Saddi Blehmeh): An indication of the strength and firmness of a relationship between two friends, so they are like the warp and weft threads woven together on a loom. The warp is the longitudinal thread of weaving, and the weft is the transverse thread.

7) O you who build a wall out of thread (Ya bani het min khet): Indicates the difficulty of brocade weaving.

Folk Songs on the Brocade Fabric

1) O loom, weave! O loom, weave! ... A gilded brocade whose threads are woven.
O teacher, sing! O teacher, dance! Our profession is the most beautiful one in the universe! [Note: The teacher in this verse is the teacher of craftsmen, the mu'allim.]

2) My darling has brought me, Brought me a brocade!
My darling's profession is the most beautiful profession.
He brought me the gilded, he brought me the veiled,
A composite brocade ... and asked me to stay up at night.
My darling has brought me, Brought me a brocade!
My darling's profession is the most beautiful profession.
He said, "O pretty!" Brocade to the pretty lady.
How beautiful the warp is! And the moon goes around.

Appendix 2: Quantitative Table of Some of the Study's Results

Before the war	After seven years of war
In general	
Number of craftsmen working in the field of traditional Damascene textile handicrafts registered in the Union of Craft Associations in Damascus: 250 in 2011	Number of craftsmen working in the field of traditional Damascene textile handicrafts registered in the Union of Craft Associations in Damascus: 65 in 2018
Sales of all handicraft products decreased by 80% to 95% from 2011 until the date of writing this paper	
Aghabani fabrics	
There were about 15,000 Aghabani machines in Douma and its environs.	It is difficult to estimate of the number of machines that still exist in Douma. It is estimated that about 1,000 machines were taken to Damascus, and are in use there almost daily.
Number of persons practising the Aghabani printing profession: 8	Number of persons practising the Aghabani printing profession: 2
Number of hours craftsmen in the Aghabani printing profession work each day: 6 to 7	Number of hours craftsmen in the Aghabani printing profession work each day: 3, due to the lack of demand
The price of a kilogram of Aghabani yarns is $5.	The price of a kilogram of Aghabani yarns is $8
The price of a new Aghabani machine is $1,000.	Today, the machines in the market are mostly used ones and priced from $300 $500; new ones are priced around $1,000
Number of persons proficient in the repair of the Aghabani machines in Douma: 10	We found one person proficient in the repair of the Aghabani machines in Damascus. There is no information on who is still proficient in machinery repairs in Douma.
Brocade	
The duration of working in silk dyeing: **throughout the week**	The duration of working in silk dyeing: **two days a week,** due to the lack of demand
Number of workers operating looms with Antoun Mouzannar, one of the most important brocade producers: 9	Number of workers operating looms with Antoun Mouzannar: 3
The price for a kilogram of silk is $**70**	The price for a kilogram of silk is $**100.**

Zeina Shahla

Shahla is a journalist residing in Damascus, Syria. She holds a Bachelor's Degree in Information Engineering from Damascus University, 2004, and a Bachelor's Degree in Media from Damascus University, 2017. She is currently preparing for a diploma in International Relations and Diplomatic Studies from the Syrian International Academy in Damascus.

She works as an independent journalist with a number of media agencies including Alquds Alarabi Newspaper, Netherlands Radio and Radio Suriyat. She is interested in the social and humanitarian stories of Syrians inside and outside Syria. Shahla also focuses on the field of investigative journalism with Arab Reporters for Investigative Journalism Network. She also participated in a number of social and cultural research papers with several independent organizations including the Al-Asfari Institute for Civil Society and Citizenship at the American University of Beirut.

ibidem.eu